JETS AT SEA

JETS AT SEA

Naval Aviation in Transition
1945–1955

by

LEO MARRIOTT

Pen & Sword
AVIATION

First published in Great Britain in 2008 by
Pen and Sword Aviation
an imprint of
Pen & Sword Books Ltd
47 Church Street
Barnsley
South Yorkshire
S70 2AS

ISBN 978-1-84415-742-6

A CIP catalogue record for this book is
available from the British Library.

Typeset in 11/13 Palatino by
Concept, Huddersfield, West Yorkshire

Printed and bound in England by
CPI UK

Pen & Sword Books Ltd incorporates the Imprints of Pen & Sword
Aviation, Pen & Sword Maritime, Pen & Sword Military,
Wharncliffe Local History, Pen & Sword Select, Pen & Sword Military
Classics, Leo Cooper, Remember When, Seaforth Publishing and
Frontline Publishing.

For a complete list of Pen & Sword titles please contact
PEN & SWORD BOOKS LIMITED
47 Church Street, Barnsley, South Yorkshire, S70 2AS, England
E-mail: enquiries@pen-and-sword.co.uk
Website: www.pen-and-sword.co.uk

CONTENTS

GLOSSARY

AA	Anti Aircraft
A&AEE	Aircraft and Armament Experimental Establishment
ACS	Aircraft Carrier Squadron
AEW	Airborne Early Warning
AI	Airborne Interception (Radar)
ASW	Anti Submarine Warfare
BuAer	US Navy Bureau of Aeronautics
CAG	Carrier Air Group
CCA	Carrier Controlled Approach
C of G	Centre of Gravity
CVF	Future Aircraft Carrier
DP	Dual Purpose
ehp	Equivalent Horse Power
eshp	Equivalent Shaft Horse Power
EW	Electronic Warfare
FB	Fighter Bomber
FR	Fighter Reconnaissance
HA	High Angle
HMS	His/Her Majesty's Ship
HMAS	His/Her Majesty's Australian Ship
HMCS	His/Her Majesty's Canadian Ship
HNMS	Her Netherlands Majesty's Ship
hp	Horse Power
HVAR	High Velocity Aerial Rocket
IAS	Indicated Air Speed
LPH	Landing Platform (Helicopter)
lb st	pounds static thrust (measure of jet engine power)
lb thrust	pounds thrust (same as pounds static thrust)
MAUW	Maximum All Up Weight
MDAP	Mutual Defence Assistance Pact
MTOW	Maximum Take Off Weight
NACA	National Advisory Committee for Aeronautics (US organisation)

NAS	Naval Air Squadron or Naval Air Station
NATO	North Atlantic Treaty Organisation
nm	Nautical Mile
OTC	Officer in Tactical Command
PR	Photographic Reconnaissance
RAE	Royal Aircraft Establishment (Farnborough)
RAF	Royal Air Force
RAN	Royal Australian Navy
RATOG	Rocket Assisted Take Off Gear
RCN	Royal Canadian Navy
RN	Royal Navy
RNN	Royal Netherlands Navy
RNVR	Royal Navy Volunteer Reserve
RNZAF	Royal New Zealand Air Force
SAR	Search and Rescue
SBAC	Society of British Aircraft Constructors
STOVL	Short Take Off Vertical Landing
TACAN	Tactical Aid to Navigation
US	United States
USAAF	United States Army Air Force
USAF	United States Air Force (officially formed 1948)
USN	United States Navy
USS	United States Ship
VTOL	Vertical Take Off and Landing

US NAVY AIRCRAFT DESIGNATIONS

During the period under review in this book the US Navy applied a system of designations that indicated the operational role of the aircraft and its manufacturer. The first letter(s) indicated the aircraft's function and the second letter the manufacturer. Between the two letters was a number that indicated the sequence of designs or projects from that particular manufacturer. Thus, for example, the well known DC-3 civil airliner became the R4D when used by the US Navy as a military transport, indicating that it was the fourth Transport design (R) produced by Douglas (D). Other numbers and letters could be added to indicate sub variants such as the R4D-5T, indicating a Mark 5 (with different engines) adapted for the Training role (T). In the case of the first design of a particular type the number one was omitted, an example being the Douglas Skyraider, which, as the first Douglas Attack aircraft, was designated AD with a subsequent number for sub variants (e.g. AD-4). One major variation that will often be noted in the text is the prefix X for a prototype and Y for a pre-production aircraft, an example being the XF6U-1 Cutlass – prototype of the sixth fighter design by Chance Vought. Relevant manufacturers and function codes are given below, although this list is not exhaustive and many minor functions and manufacturers are omitted for the sake of brevity. This basic system remained in use until 1962 when a common unified designation system was introduced for all US military aircraft (Army, Navy and Air Force).

Type Designators

A   Attack (from 1946 onwards)
B   Bomber

BT	Bomber – Torpedo (changed to A in 1946)
F	Fighter
H	Helicopter (normally with additional suffix)
J	Utility
N	Trainer
O	Observation
P	Patrol
PB	Patrol Bomber
R	Transport
S	Anti Submarine (from 1951)
S	Scout (to 1946)
SB	Scout Bomber (to 1946)
SO	Scout Observation (to 1946)
T	Trainer (from 1948)
TB	Torpedo Bomber (to 1946)
U	Unpiloted Drone (from 1946)
W	Electronic Search (from 1952)

Manufacturers' Codes

(Note: the same letter was often allocated to more than one manufacturer.)

B	Boeing
C	Curtiss
D	Douglas
D	McDonnell (later changed to H)
F	Grumman
G	Goodyear
K	Kaiser-Fleetwings
M	Glenn Martin
M	General Motors (Eastern Aircraft Division)
N	Naval Aircraft Factory
O	Lockheed
R	Ryan
S	Sikorsky
U	Chance Vought
V	Lockheed (Vega Plant)
Y	Convair (previously Consolidated)

INTRODUCTION:
A BRIEF HISTORY OF NAVAL
AVIATION TO 1945

In the modern world the most powerful navy is that of the United States, a mantle it inherited from Britain's Royal Navy in the course of World War II. At the heart of its ability to support the nation's political policy with military force deployed flexibly and promptly, is its fleet of large aircraft carriers, each carrying an air wing of over eighty aircraft. These range from agile, high-performance fighters to strike aircraft capable of delivering massive ordnance loads with pinpoint accuracy hundreds of miles away from the parent ship. These aircraft are every bit as capable and potent as their land-based counterparts and make no concessions to the fact that they operate from the relatively restricted confines of a carrier flight deck. This is a situation that has now existed for almost fifty years and the techniques employed to safely operate such aircraft have only changed in detail over that period. Heavily laden aircraft are thrown into the air by powerful but smoothly accelerating steam catapults (even aboard the nuclear-powered carriers) and on returning to the ship the pilot is guided home by electronic navigation aids, making a safe approach with visual cues supplied by a stabilised mirror landing sight. As he touches down the aircraft's hook engages the arrester wires and it is brought to a rapid stop. However, in the event of missing the wires, the pilot only has to guide the aircraft along the offset angled flight deck, engines already set at full power on touchdown just in case of such an eventuality, and lift off for a further circuit and landing.

Compare this with the state of the art in 1945. Aircraft were launched by less powerful hydraulic catapults with harsh acceleration or, if heavily laden, made a rolling take-off along the full length of the fore and aft orientated flight deck. Although some electronic navigation aids were available to guide the pilot back to the ship, the final approach was carried out under the guidance of batsmen who stood at the port edge of the flight deck and attempted to give indications to the pilot by means of coloured or illuminated bats. This required an enormous amount of teamwork and trust between those involved. If the aircraft missed the arrester wires the inevitable result was an engagement with the wire and webbing crash barrier whose prime function was to stop the landing aircraft running into other aircraft parked forward, rather than to save the aircraft or pilot. In the worst case scenario the landing aircraft might bounce over the barrier or stall after an unsuccessful attempt to go around from a baulked landing and crash into the deck park.

The change in these strikingly different methods of operation occurred in the decade immediately after World War II and by the late 1950s new carriers were entering service with all the equipment required for modern carrier flying. The pace of change was forced by the new aircraft themselves, which as well as growing ever larger and heavier, were also powered by the then still revolutionary jet engine. The characteristics of these new aircraft were entirely different from their propeller-driven predecessors and the learning curve for the British and American navies was steep and difficult. This book tells the story of this dramatic decade, which had a backdrop of the onset of the so-called Cold War with the nightmare threat of atomic warfare and a real hot shooting war in Korea where carrier aviation played a vital role and naval jet aircraft were used in combat for the first time. The coming of the jets also coincided with the ultimate developments of piston-engined aircraft and the two operated side by side during much of the period under review.

The story will be told with reference to three main strands. First a look at the ships themselves and how they were adapted to operate the new breeds of combat aircraft, then a review of the last generation of piston-engined aircraft (some of which stood up well in comparison with the early jets), and last a look at how the jet engine evolved and was applied to the early generations of naval jet combat aircraft. Although the main narrative will inevitably concentrate on the British and American navies who between them were responsible for all the major steps forward, it should not be forgotten that other navies operated aircraft carriers during the period 1945 to 1955 and benefited from the advances described in this book. However, the only one to produce indigenous naval aircraft, at least in prototype form, was the French Navy, which today operates a modern nuclear-powered aircraft carrier, the *Charles de Gaulle*.

Before looking at the post 1945 developments, it is perhaps relevant to look briefly at the history of naval aviation, which began within a decade of the Wright brothers' first successful powered flight at Kitty Hill in December 1903. As early as 1908 the US Navy was actively examining the possibility of using aeroplanes for observation and scouting missions. The Royal Navy was also aware of the potential advantages of aerial observation but initially put its efforts into developing airships and some of these did prove useful in carrying out anti-submarine patrols in World War One. In 1909 Louis Blériot made his famous flight across the English Channel, graphically illustrating that a potential enemy now had a new way of reaching and attacking the British mainland. As far as ship borne naval aviation was concerned, the most significant event took place on the other side of the Atlantic in the form of Eugene Ely's brave flight from the foredeck of the cruiser USS *Birmingham*. Subsequently he made a further flight to land aboard the USS *Pennsylvania* off San Francisco, and later made the return flight to the shore. British developments began in 1911 when Lt Samson became the first British pilot to fly off the deck of a ship in January 1912. At the time the ship was at anchor as in the American experiments but in May 1912 Samson flew a Short S.38 floatplane off the bows of the battleship HMS *Hibernia* while the ship was underway. After the outbreak of war in 1914, the Royal Naval Air Service (RNAS) expanded rapidly and seaplanes were later regularly carried aboard battleships and cruisers for scouting purposes and also for defence against Zeppelin attacks.

A more significant line of development was the conversion of several merchant ships, mostly cross Channel ferries, to act as seaplane carriers. Typically, these could carry between three and six aircraft, which generally took off from a platform over the bows and on return landed in the sea alongside, to be hoisted aboard by cranes. With modifications, the aircraft complement was slightly increased later in the war and on the largest and best known, HMS *Campania*, eventually carried up to eleven seaplanes. This method of operation was cumbersome and severely restricted tactical flexibility, while the ships themselves were slow compared with regular warships. In 1917 the large cruiser HMS *Furious*, then still under construction, was altered by the removal of the forward 18 inch gun turret, which was replaced with a hangar and flying-off deck. After the ship commissioned, Squadron Commander EH Dunning experimented with a Sopwith Pup fighter fitted with a normal wheeled undercarriage. Flying off from the forward flight deck, he flew round the ship and approached up the port side before sideslipping back onto the deck where handlers were ready to grasp specially fitted toggles attached to the wingtips. This was the first time that such a feat had been accomplished and it paved the way for carrier aviation as we know it today. Unfortunately, Dunning was killed only a few days later when he experienced an engine failure while carrying out further trials. Nevertheless it was already realised that his pioneering method was not suitable for normal operational use and *Furious* was withdrawn for further modifications. She emerged from refit in 1918 with her after gun turret also removed and again this was replaced with a hangar, the roof of which constituted a flying on deck. Although this was a great improvement it was not the complete answer. This arrived in the shape of HMS *Argus*, the world's first proper aircraft carrier with a full-length unobstructed flight deck, which had been converted from an uncompleted Italian liner, *Conte Rosso*. She was equipped with a balanced air group consisting of eight Sopwith Camel fighters and twelve Sopwith Cuckoo torpedo bombers – the sort of numbers that the Royal Navy subsequently struggled to achieve in the interwar years due to the fact that control of the squadrons and their aircraft was naively passed to the fledgling Royal Air Force in 1918.

Despite this, the Royal Navy commissioned several carriers in the following decade, including the rebuilt *Furious* and her sister ships, *Courageous* and *Glorious*, the latter pair capable of accommodating up to forty-eight aircraft.

Another large carrier was HMS *Eagle*, converted from an ex-Chilian battleship and distinguishable by its twin funnels. In 1923 the world's first aircraft carrier which had been designed from scratch for the role commissioned in the shape of HMS *Hermes*, although she was considerably smaller than the others. Thus by 1930 the Royal Navy had no fewer than six carriers in service while the rival United States Navy could muster only three, one of which was a converted collier, although the other two, *Lexington* and *Saratoga*, were 33,000-ton converted battle-cruisers, which were to give sterling service in World War II.

As the threat of war loomed in the 1930s, Britain's naval re-armament began to accelerate. An early project was the laying down of the 22,000-ton aircraft carrier *Ark Royal* in 1934, which was completed in 1938 and was capable of carrying up to seventy-two aircraft. She was followed by four Illustrious class carriers, of which the first three commissioned in 1940/41, and the slightly modified *Indomitable* also followed in 1941. Initially these were designed to operate thirty-six or forty-eight aircraft, although wartime modifications eventually boosted this to fifty-four. Although similar in size to the earlier *Ark Royal*, the smaller aircraft complement was explained by the fact that these ships were the first carriers to incorporate an armoured flight deck, an innovation that saved them from being sunk in action on more than one occasion. *Illustrious* was heavily hit by German dive bombers in the Mediterranean in June 1941 and was out of action for over a year while both *Formidable* and *Indomitable* were hit by kamikaze attacks in the Pacific in 1945. The latter were only briefly out of action before resuming fly-ing operations and remaining on station. Two further carriers, *Implacable* and *Indefatigable*, were commissioned in 1944 and could initially operate fifty-four aircraft (later raised to eighty-one), this increase being due to the incorporation of two hangar decks.

Unfortunately, this build up of the Royal Navy's carrier fleet in the period between the wars was not matched by the

development and procurement of effective aircraft in anything
like the numbers required. The reason for this was that control
of the Fleet Air Arm was, as already mentioned, in the hands of
the Royal Air Force during this vital period. While the junior
service took its responsibilities very seriously and attempted to
build on the high standards set by the RNAS during World
War I, there was nevertheless a lack of clear thinking and
direction in respect of formulating the precise requirements
and specifications for naval aircraft. Even when the Royal Navy
eventually regained control of the Fleet Air Arm in 1937, it was
too late to have any major impact immediately and the service
went to war with a motley collection of aircraft that fell well
below the performance of similar land-based types. Its main
torpedo strike aircraft was the Fairey Swordfish biplane, which
had a top speed of less than 100 knots when carrying a torpedo.
Nevertheless, such was the spirit of the Royal Navy aircrews
that even with such limited equipment they achieved some
spectacular successes – notably the attack on Taranto. This
crippled the Italian battle fleet in November 1940 and sent out a
strong message that carrier-based aviation was now the arbiter
of naval warfare. In May 1941 it was Swordfish from the carrier
Ark Royal that succeeded in crippling the German battleship
Bismarck so that the pursuing surface forces were able to inter-
cept and sink her.

Despite Britain's initial lead in terms of carrier numbers,
the US Navy had retained its own air arm and was able to
develop aircraft, procedures and tactical doctrine using the
two carriers *Lexington* and *Saratoga* from 1927 onwards. These
imposing ships were characterised by a single massive funnel,
which was separated from the main island superstructure,
and their cavernous hangars and large flight deck enabled them
to operate up to eighty aircraft. In terms of displacement they
remained the largest carriers ever built until overtaken by the
Midway class, which were not completed until after World
War II. In the 1930s America continued a steady, if modest by
later standards, programme of carrier construction starting with
the 14,500-ton USS *Ranger*, which was completed in 1934. This
was the US Navy's first purpose-designed carrier and it could
carry up to seventy-six aircraft despite being less than half

the *Lexington*'s displacement. She was followed by *Yorktown* and *Enterprise*, which could operate around eighty aircraft on a displacement of 20,000 tons and the smaller 14,500-ton USS *Wasp*, which, despite a reduction in size due to Treaty tonnage limitations, could still operate the same number of aircraft. The last pre-war carrier to commission was the *Hornet*, which was basically a repeat of the *Yorktown* and was laid down after the Treaty restrictions had expired. By the time of the attack on Pearl Harbor the US Navy could boast a force of seven fleet carriers approximately equal to that of the Royal Navy at the time. However, the vital difference was in the numbers and types of aircraft, flown and operated by a cadre of experienced and well trained aircrew. Strike aircraft mainly comprised the Devastator and Dauntless, all-metal monoplanes, backed up by the sturdy Grumman F4F Wildcat fighter. These, and subsequent types, were all designed and built specifically for naval use and incorporated the experience gained by intensive flying operations and exercises over the previous decade.

By December 1941, the war in Europe had been raging for almost two years and hostilities with Japan had been anticipated to some degree. With all naval Treaty restrictions now no longer applicable, the US Navy was able to start the construction of a new class of large carriers, which would ultimately form the core of the great naval task forces that were to be the deciding factor in the Pacific War. The first of the new 27,000-ton Essex class carriers was laid down in April 1941 and commissioned in December 1942, a remarkably short construction time for such a large and complex ship. A further sixteen were completed before the end of World War II, while a total of twenty-four were eventually built. The nominal capacity of these ships was eighty aircraft, more or less the same as the preceding classes, but these figures disguised the fact that the Essex class would be able to operate the larger and heavier naval aircraft that were then about to enter service. Unlike the contemporary British carriers, they did not have armoured flight decks and consequently several were seriously damaged and put out of action for considerable periods when subjected to attack by Japanese kamikaze aircraft, although none were actually sunk.

The early war years saw the loss of many of the pre-war carriers, including *Lexington, Yorktown, Hornet* and *Wasp*, and they could not immediately be replaced by the Essex class then still under construction. In anticipation of a potential shortage of carriers, an emergency programme was instituted in 1941 to convert several Cleveland class cruiser hulls, then under construction, to light aircraft carriers. The original hull and machinery were retained, a hangar deck was built onto the hull and this was the covered by a narrow flight deck. Funnel uptakes were angled to starboard and a small island superstructure was mounted forward of these. The resulting Independence class eventually numbered some nine ships and, crucially, all commissioned before the end of 1943. The design was obviously something of a compromise and only thirty aircraft could be carried for operational purposes, although when employed in the aircraft ferry role, they could stow up to 100. Their great virtue was a speed of 32 knots, which allowed them to take their place in the fast carrier task groups forming for the great sweep across the Central Pacific.

The last major US carriers laid down in World War II were the three Midway class, which now followed British practice with significant armour protection to both the hull and flight deck. On a full load displacement of around 60,000 tons, they would be capable of carrying up to 137 aircraft. However, although work began in 1943, none of the three laid down were to be in service before the end of the War and their subsequent story is described in Chapter 3.

During World War II, Germany and Italy both attempted to build and commission aircraft carriers but none were completed. Consequently, the initial role of the Royal Navy carriers was to operate in conjunction with surface ships to locate and run down German surface raiders and also to support the abortive land operations in Norway. Early losses included HMS *Courageous*, which was torpedoed while hunting for submarines, and HMS *Glorious*, which was sunk by gunfire from the battlecruisers *Scharnhorst* and *Gneisenau* off Norway. In the Mediterranean there were some successes, such as Taranto, and HMS *Formidable* played an important part in the Battle of Matapan. Subsequently *Illustrious* was badly damaged by air attack in June 1941 and was out of action for almost a year, while *Formidable* was also out of action from May 1941 for six months. The much loved *Ark Royal* was torpedoed and sunk in November 1941. Nevertheless, the Royal Navy was able to field no fewer than four carriers (*Indomitable, Victorious, Eagle* and *Argus*) in support of Operation *Pedestal*, a valiant attempt to resupply the vital island of Malta in August 1942. Despite grievous losses, enough ships got though to keep the island going although *Eagle* was sunk and *Indomitable* was severely damaged. However, this action marked the turning of the tide and by the end of the year German forces were being swept out of North Africa and in 1943 Sicily and then Italy were successfully invaded. This released the battered carriers and their worn out air groups for much needed refits and re-equipping before being sent out to the Far East where they were eventually able to join with their American cousins for the final push against Japan.

In the meantime the US Navy had suffered setbacks of its own before eventually turning the tide of the Pacific War in the Battle of Midway in June 1942. Prior to that the USS *Lexington* had been lost in the Battle of the Coral Sea in May 1942, the first time that the action had been entirely between the two carrier air fleets with ships never coming in visual contact with their opponents. At Midway, US naval aircraft sank four Japanese carriers for the loss of the *Yorktown*, although the USS *Wasp* was subsequently lost in operations around Guadalcanal in September 1942 and the USS *Hornet* was sunk at the Battle of Santa Cruz in the following month. The remaining carriers, *Saratoga* and *Enterprise*, were both damaged and undergoing repair by the end of 1942 and to help hold the ring the British carrier HMS *Victorious* was dispatched to the Pacific, teaming up with the USS *Saratoga* for a few months in 1943. From that time onwards the new Essex class rapidly passed into service, backed up by the smaller Independence class so that the push across the Pacific could begin.

New aircraft were also coming into service, including the Grumman F6F Hellcat and Chance Vought F4U Corsair fighters, while the Grumman Avenger torpedo bomber and Curtiss Helldiver dive bomber made up the strike forces. The first three of these were supplied in considerable numbers to the Royal Navy where they largely replaced British-designed aircraft aboard the fleet carriers. By August 1945 the only British-made aircraft deployed in significant numbers with the British Pacific Fleet was the Seafire Mk.III and there were also a few squadrons of the Fairey Firefly strike fighter, which had made its combat debut over Norway in 1944. When the Pacific War ended in August 1945, almost the entire Royal Navy carrier fleet was deployed in the Far East, including five out of six fleet carriers (*Illustrious* was at home refitting) and four new light fleet carriers of the 11th ACS (*Venerable, Colossus, Glory* and *Vengeance*) whose air groups each comprised twenty-one Corsair fighters and eighteen Barracuda torpedo bombers. The fleet carriers made up the 1st ACS and became designated Task Force 37 with approximately 255 aircraft embarked. However, it is an indication of the relative status of the two navies at this stage of the war when it is realised that the US Navy Task Force 38 comprised over 1,200 aircraft embarked on no fewer than eight Essex class and six Independence class carriers.

1

PROPELLER SWANSONG – ROYAL NAVY

By the end of World War II the future of combat aircraft clearly lay with jet propulsion but it was to be several years before this new generation would reach a maturity that would enable them to be deployed in significant numbers at sea. In the meantime piston-engined fighters and strike aircraft remained in service and several new types, too late to see war service, were introduced. Some of these represented the ultimate development of the propeller-driven aircraft and had performance figures not far removed from those of the first jets. The great difference, of course, was that whereas the jets were only at the beginning of their development, the prop aircraft had more or less reached the limits of what was possible with their form of propulsion. Nevertheless, several piston-engined fighters and multi-role naval aircraft remained in front-line service almost to the end of the decade covered by this book. In the case of the Royal Navy, considerable effort was expended in their development as described in the following pages.

In the closing stages of World War II the standard British naval fighter was the Supermarine Seafire Mk.III, which had been developed from the land-based Spitfire Mk.VC. This was the first Seafire version to feature wing folding, although this was done manually, and a total of 1,263 were built and delivered from June 1943 onwards. By mid 1945 some eight squadrons of Seafire IIIs were deployed to the Far East for use aboard the carriers *Indefatigable* and *Implacable* of the British Pacific Fleet for the final operations against Japan. These two ships each had two hangar decks, although the headroom was reduced to only 14 feet (compared with 16 feet in the earlier *Illustrious* class) so that American aircraft with upward-folding wings such as the Corsair would not fit and only the Seafire could be accommodated. However, even by 1945 the Seafire Mk.III was bordering on the obsolescent and its performance, both in terms of speed and range, was generally well below that of the contemporary American-built Vought Corsair, which was used in large numbers by the British Pacific Fleet aboard its other carriers. The Seafire's great redeeming feature was its performance at low levels and its blistering rate of climb up to medium altitudes, making it ideal for its primary role as an interceptor. But it lacked the load-carrying ability and range of the American aircraft and this made it less useful in offensive operations against the Japanese mainland. However, the Royal Navy Corsairs and Hellcats were all supplied on Lend-Lease and consequently had to be discarded when peace came, leaving the Seafire to soldier on. After the War, the Mk.III was quickly withdrawn from service and twelve were sold to the Irish Air Corps in 1947 while another forty-eight were transferred to the French Navy.

Fortunately by mid 1945 a better-performing Seafire was just coming into service in the shape of the Griffon-powered Mk.XV and this was basically similar to the RAF's Mk.XII Spitfire. The naval version utilised a Mk.III airframe modified to accept a 1,750 hp Mk.VI Griffon engine, which naturally boosted performance, including an increase in maximum level flight speed from 359 to 392 mph, while the maximum rate of climb was now in excess of 4,000 feet/min. Initial production examples had an under fuselage arrester hook carried on an A-frame, which pivoted just aft of the wing trailing edge. However, later aircraft had a new sting-type arrester hook, which could also pivot laterally, and this was anchored to the base of the fuselage sternpost, which also carried the rudder. The incorporation of this resulted in a change of the rudder profile, making it slightly taller.

The first unit to equip with the Seafire XV was 802 Squadron based at Arbroath in May 1945. Later in the year examples were ferried out to the Far East to equip 801 Squadron, which was disembarked from HMS *Implacable* at RAN air station Schofields at the end of 1945. At this time the new aircraft was not cleared for deck operations and consequently was not embarked while the ship was in Australian waters. At home 805 and 806 Squadrons also received Seafire XVs in January 1946 and in that year the type was deployed aboard the light fleet carriers *Venerable*, *Glory* and *Ocean*, replacing the Lend-Lease Corsairs and Hellcats previously embarked. The Seafire XV was also selected by the Royal Canadian Navy to equip a planned force of two light fleet carriers. Two squadrons, 803 and 883, were formed in the late summer of 1945, although the latter was disbanded in February 1946 and only 803 Squadron was embarked aboard HMCS *Warrior* when she sailed for Canada in March 1946.

This increasing deployment of the Griffon-engined Seafire was suddenly halted in mid 1946 when it became apparent that there were major problems with the new engines when operated in the naval environment. These centred on the engine's M ratio supercharger clutch, which was prone to slipping at high rpm and high manifold pressure – the very conditions used for carrier take-offs, and wave offs from baulked landings. Consequently the type was withdrawn from carrier operations until early 1947 when the problem was eventually overcome and the necessary modifications incorporated. This episode had unfortunate ramifications for the front-line squadrons operating from the light fleet carriers, which had been organised into Carrier Air Groups (CAG), each consisting of one squadron of Seafires and one of Fireflies. As a temporary expedient, some of the Seafire squadrons were equipped with Fireflies.

Immediately after the end of the War, in September 1945, yet another Seafire variant began joining the fleet. This was the Seafire F.XVII, which was a progressive development of the Mk.XV. The most obvious external difference was the adoption of a cut-down rear fuselage and the fitting of a clear-view bubble canopy (this modification had been extended to some of the last Mk.XVs built by Westlands). Other improvements included a strengthened rear spar (which permitted the carriage of wing-mounted stores), a 24-volt electrical system (instead of 12-volt) and provision for RATOG gear. Another important feature was the adoption of a long stroke undercarriage, which was better able to absorb the shocks imposed by a carrier landing and permitted American-style deck landings where the aircraft was flown straight onto the deck rather than flared for a more traditional, and softer, landing. Earlier Spitfires with unmodified undercarriages originally only intended for land-based operations were notoriously prone to bounce on landing with a resulting risk of missing the arrester wires and ending up in the crash barrier or, even worse, missing the barrier and hitting aircraft in the forward deck park.

Once the problems with the Griffon supercharger were solved, the Mk.XVII supplemented and eventually replaced the Mk.XV in the front-line squadrons, and a total of 232 of this variant were built (compared with a total of 384 Seafire XVs). In the summer of 1947 the first RNVR squadrons were formed and all of these (1830, 1831, 1832 and 1833 Squadrons) operated Seafire XVIIs at various times. With the introduction of newer types the Seafire XVII was eventually relegated to the training role and the Royal Navy's last Seafire unit was 764 Advanced Training Squadron based at Yeovilton, which was still flying Mk.XVIIs until it was disbanded in November 1954.

Despite its longevity, the Mk.XVII was not the end of the Seafire development line. The basic Spitfire design underwent a continuous process of evolution almost from the day of the flight of the first prototype in 1936 and the pace of such changes accelerated rapidly under the stimulus of war conditions. To improve performance more powerful versions of the original Rolls-Royce Merlin were fitted, and these in turn gave way to the later Rolls-Royce Griffon. Armament was increased, constant speed propellers of three, then four and five blades were fitted to absorb the increased power, while the steady addition of new military equipment inexorably drove up the weight of the aircraft. It was recognised at an early stage that a strengthened wing and undercarriage would be needed to cope with this but while Spitfires were urgently needed in all the war theatres there was neither the time nor the resources to implement such a change on the production lines. Eventually, following trials with a single Spitfire Mk.XX prototype, the definitive F.21 entered production in 1944, production deliveries to the RAF commencing in September of that year. The Spitfire F.21 featured the new much-strengthened wing fitted with new ailerons, which together much improved manoeuvrability, particularly at the higher altitudes. The standard armament was now four 20 mm Hispano cannon and a lengthened and stronger undercarriage was fitted. This featured larger wheels and tyres, necessitating a slight bulge in the outer undercarriage doors, which now completely covered the wheels when retracted. The well known elliptical outline was retained, but production aircraft had the tips rounded off so that the overall span was increased by only an inch.

The Admiralty had naturally kept a close eye on these developments and a naval version designated Seafire F.45 was ordered under Specification N.7/44. The prototype was in fact a converted F.21 (TM379) modified to naval standards by Cunliffe-Owen. The most obvious change was the addition of a stinger-type arrester hook faired under the lower edge of the rudder. However, no wing folding was provided, limiting use of production F.45s to shore-based units, the first being 778 Squadron in November 1946 – two years after the original specification was issued. Subsequently some aircraft were fitted with cameras and designated FR.45. One potential problem was the enormous torque generated by the 2,035 hp Griffon 61 engine and rudder area was increased to help counteract this on take-off. Although not used operationally at sea, carrier trials were conducted aboard HMS *Pretoria Castle* and some 239 landings were made with only minor incidents. One Seafire F.45 was tested with a six-bladed contra-rotating propeller and two more were fitted with the more powerful Griffon 85, also driving a contra-rotating propeller. The success of these installations led to the Seafire FR.46. Although the first few were completed with five-bladed single propellers, production eventually reverted to the contra-rotating installation. The FR.46 was basically a naval equivalent of the RAF's Spitfire F.22 and featured the cut-down rear fuselage and tear drop canopy, although it was still regarded as an interim type with no provision for wing folding. Only twenty-four were produced and they mostly went to training and RNVR units (781 and 1832 Squadrons), although three were allocated to the Empire Test Pilots' School.

The final Seafire variant, and in fact the very last of the famous Spitfire line, was the FR.47, which finally incorporated wing folding, the hinge being outboard of the guns. Initially it was done manually with a jury strut to secure the wings when folded but eventually hydraulic power folding was incorporated and the strut was no longer necessary. With wings folded the overall height was only 13 feet 10 inches, well within the hangar capacity of all British carriers including the light fleets, and the folded span was 25 feet 5 inches. The main armament of four 20 mm Hispano cannon was carried but these were a new short-barrelled version, which provided a recognition feature to distinguish it from the earlier FR.46. In theory three 500 lb bombs could be carried, one under the centre fuselage and one under each wing, but in practice it was rare to carry more than two and these were quite often the lighter 250 lb bombs. Both versions had the taller fin and rudder assembly derived from that fitted to the contemporary Spiteful/Seafang (described later). The FR.46 and FR.47 both carried two heated cameras, one vertical and one oblique, and these were the reason for the FR (fighter reconnaissance) designation. One distinguishing feature of the FR.47 was the chin intake of the carburettor immediately behind the propeller spinner, which resulted in a cleaner nose profile. However, it was discovered that wash from the contra-rotating propellers interfered with the airflow through the intake and reduced engine performance. Consequently later production aircraft had the carburettor intake moved back to the traditional position under the rear of the engine cowling almost in line with the wing leading edge. In all, a total of ninety Seafire 47s were produced, the last being delivered in March 1949. This was the very last of the illustrious Spitfire line as the RAF's last Spitfire F.24 had been delivered in February 1948, a total of over 22,000 Spitfire/Seafires having been produced since 1936.

Despite coming on the scene when the emphasis was beginning to turn to jet combat aircraft, the Seafire 47 had a very respectable performance. Apart from a top speed of 451 mph at 20,000 feet, it had an initial rate of climb of 4,800 feet/min and a service ceiling of over 43,000 feet. These were very creditable figures matched by very few other production piston-engined fighters and barely matched by some of the first-generation jets. An interceptor for fleet defence, the Seafire was still an effective aircraft but its potential for offensive use was limited by its restricted range when carrying any significant payload. In addition, the original Spitfire airframe had never been intended for carrier use and despite constant improvements and strengthening, there were always problems with the strains caused by arrested landings, as well as handling difficulties associated with the outward retracting narrow track undercarriage. Admittedly, the introduction of contra-rotating propellers on the last-generation Seafires considerably eased some of the problems associated with flying from carrier decks.

The FR.47 entered service in February 1948, equipping 804 Squadron, which was initially based at RNAS Ford but subsequently embarked on the light fleet carrier HMS *Glory*. The only other frontline squadron to receive the new version was 800 Squadron at Donibristle in April 1949 prior to embarking in the HMS *Triumph* for a deployment to the Far East. Here 800 Squadron's Seafire 47s became the only examples of the post-war variants to see action, initially in support of British troops fighting communist guerrillas in Malaya during late 1949 and early 1950. A sterner test came with the outbreak of the Korean War where *Theseus* and her air group of Seafires and Fireflies were heavily committed from the outbreak of hostilities in June 1950 until the following September. By this time her battered air group had an effective strength of only eleven aircraft. Losses from enemy action were few but attrition and accidents had taken their toll, including overstressing of some Seafire rear fuselages. When *Triumph* returned to the UK in November 1950, the Seafires were disembarked and the frontline career of the Seafire was ended. However, others continued to serve with second-line units and the last flying examples served with 1833 RNVR squadron at Bramcote until 1952.

The continued development of the Spitfire and Seafire in World War II had brought the problems of high-speed flight into sharp focus. One reason for the longevity of the basic Spitfire design was RJ Mitchell's foresight in specifying and designing a very thin wing with a thickness/chord ratio of only 9 per cent at the tips. This was substantially less than its early contemporaries and during and after the War some Farnborough-based Spitfires achieved velocities in excess of Mach 0.9 in a series of trials involving vertical dives. However, these were not without excitement and drama as well as significant damage in some cases, including a lost propeller in one instance. Interestingly, similar trials with Mustangs and Thunderbolts failed to achieve such high speeds due to their much-increased drag at high speeds. However, aerodynamic design had advanced considerably since the mid 1930s when the Spitfire was first designed and in particular the benefits of a laminar flow wing with maximum thickness set further back, at around 40 per cent main chord, were becoming apparent in the context of high-speed flight. Accordingly Supermarine began work on the design of such a wing in late 1942 with the intention of marrying it to the Spitfire airframe. It was hoped that such a combination would result in a faster aircraft with improved manoeuvrability and the Air Ministry issued Specification F.1/43 for a fighter equipped with a laminar flow wing, ordering three prototypes from Supermarine. These were given the Type number 371 and named Spiteful in recognition that the overall changes had resulted in a new aircraft. From the start provision was made for a naval version and the two-spar wing was designed such that a wing-folding mechanism could be incorporated if required. Another feature that had positive implications for naval use was a complete redesign of the undercarriage, which now retracted inwards and increased the wheel track by some four feet. The Seafire had always suffered from its narrow track undercarriage, which was never intended for carrier use.

In 1944 the laminar flow wing was fitted to a modified Spitfire Mk.XIV (NN660) and this aircraft first flew on 30 June, piloted by Jeffrey Quill, Supermarine's Chief Test Pilot. Unfortunately this aircraft was destroyed in an accident on 13 September 1944 and the first production standard Spiteful did not fly until 8 January 1945. Flight testing revealed that although the new wing conferred some benefits at high speed, there were significant handling problems at low speeds and a considerable amount of work was required to rectify this situation. Despite this, initial orders were placed for 188 Spitefuls, these being either F.14s with a Griffon 69 engine driving a five-blade Rotol propeller or F.15s with a Griffon 89 or 90 and a six-blade contra-rotating Rotol propeller. A third version was the Spiteful F.16 with a Griffon 101 driving a Rotol five-blade propeller. The sole example of this version is recorded as having reached a speed of 494 mph in level flight while on test from Boscombe Down in 1947. This was the highest level speed ever recorded by a British propeller-driven aircraft and was only exceeded by the American experimental XP-47J Thunderbolt, which attained 504 mph as early as August 1944. Despite this level of performance, only seventeen Spitefuls were completed before all production contracts were cancelled at the end of the War, and not all of these were flown.

As already related, the Royal Navy was being equipped with the Seafire XV at the end of the War and as early as October 1943 Supermarine had proposed fitting the new laminar flow wing to this version, the resulting aircraft being designated Type 382. However, at the time the Admiralty preferred to prioritise other projects and it was not until early 1945 that Specification N.5/45 was issued to cover production of a navalised Spiteful to be known as the Seafang, orders being placed in May for 150 aircraft. An interim prototype was quickly produced by fitting a Spiteful F.14 with a sting arrester hook for carrier trials but the true Seafang 32 prototype did not fly until June 1946. This fully navalised aircraft had hydraulic wing folding and was powered by the Griffon 89 engine driving a six-bladed contra-rotating propeller. A distinguishing external feature was the deeper engine cowling with the lip of the air intake immediately below the propeller hub, as in the Seafire 47. As with the Spiteful, the centre fuselage was slightly deeper than the standard Spitfire/Seafire, which allowed the cockpit to be raised in order to improve the pilot's view over the nose. The result of all these changes was to produce an aircraft that was much more suited to carrier operations than the basic Seafire. Test flights and deck landing trials (aboard HMS *Illustrious* in May 1947) confirmed that the pilot's view of the flight deck on approach was very good, the contra-rotating propeller eliminated issues connected with torque on take-off or in the event of a wave off, and the wide track undercarriage with long stroke oleos was much better suited to the carrier environment.

Despite the success of these trials the Seafang was not ordered in quantity, the Admiralty preferring to stick with developments of the Griffon-engined Seafire pending the introduction of jet fighters, which were by now a very realistic prospect. In all, ten interim Seafang 31s were delivered, these being fitted with an arrester hook and long stroke undercarriage, but no folding wing mechanism, while a further eight fully navalised Seafang F.32s were delivered. Of the latter only two were actually completed and flown, the others being in dismantled condition. Nevertheless the Spiteful and Seafang represented the ultimate British piston-engined fighter designs and the experience gained with the laminar flow wing was of crucial importance in developing the first generation of naval jet fighters.

While Supermarine persevered in the post-war years with Seafire and Seafang developments powered by the liquid-cooled in-line Rolls-Royce Griffon engine, the rival Hawker company was working on a new naval fighter powered by the Bristol Centaurus air-cooled radial engine. This was the Sea Fury, which eventually entered service in 1947 and went on to entirely replace the Seafire in frontline squadrons. The origins of the Sea Fury could be traced back to the Typhoon and Tempest fighters developed for the RAF, which in turn had come from Specification F.18/37 for a twelve-gun interceptor powered by a new generation of twenty-four cylinder liquid-cooled in-line engines then becoming available. One aircraft powered by a Rolls-Royce Vulture engine flew as early as October 1939 and this version was named the Tornado. Although the various prototypes provided valuable data and acted as test beds for a variety of other engines, the Tornado never entered service due to problems with the Vulture engine. However, the Napier Sabre-powered version, the Typhoon, was more success-ful (although even the Sabre engine itself was not without problems), and it was eventually produced in large numbers for the RAF. Entering service in the summer of 1942, as an interceptor it was not an outstanding success despite a top speed of over 400 mph. However, it was later to find its true role as a ground attack and close support aircraft where its good low-level performance and hard-hitting armament of four 20 mm cannon and eight 60 lb rockets or two 500 lb bombs wrought havoc amongst German ground forces, particularly in the battles following the Normandy invasion in June 1944.

Trials with the early Tornado and Typhoon prototypes led to a proposal for a version of the latter with a new thinner wing of elliptical planform and this together with other changes and improvements resulted in a new type known as the Tempest. The original design, the Tempest Mk.I, was to be powered by a Napier Sabre IV cooled by radiatiors inset into the inboard wing leading edges. Other versions were planned with Centaurus, Griffon or Sabre II engines and in fact it was the Tempest Mk.V powered by the latter that was the first to fly, in September 1942. In this version the twenty-four cylinder Sabre engine was cooled by an annular radiatior in a prominent housing under the nose as on the Typhoon. Again produced in large numbers, the Tempest entered RAF squadron service in April 1944 and subsequently played an important role in the closing stages of the war in North West Europe. The Centaurus radial-engined version eventually saw service as the Mk.II but was not operational until 1946.

By the standards of 1939/40 when Typhoon and Tempest development began to gather pace, the airframe and engine combination was much larger and heavier than the contemporary Spitfires and Hurricanes. Accordingly Hawkers later reviewed the design and proposed a so-called Tempest Light Fighter in which the wing centre section was deleted in order to reduce span, and various other changes were made throughout the airframe. Specification F.2/43 was issued to cover this design and work commenced to build prototypes of the aircraft intended for service with the RAF. The power unit was to be a Bristol Centaurus XII air-cooled radial engine and the installation of this was to be much influenced by experience gained in testing a captured Focke-Wulfe FW.190 in 1942. At the time the German fighter was markedly superior to the Spitfire Mk.V and barely matched by the Mk.IX, which was rushed into service as a stopgap. In 1943 the Admiralty issued Specification N.7/43 and Hawker suggested that this could be met by the F.2/43 design with modifications for naval service. This proposal was accepted and development of both versions proceeded apace with responsibility for naval modifications being delegated to Boulton Paul Aircraft Ltd. The prototype F.2/43 flew in September 1944 powered by a Centaurus XII but other prototypes were subsequently flown with Griffon and Sabre engines (one of the latter achieved a speed of 485 mph in 1947). By the end of 1944 the name Fury had been adopted for the RAF version and the naval variant naturally became the Sea Fury.

Even before the first flight, orders for 200 Furies and 200 Sea Furies were placed in April 1944. However, the RAF orders were cancelled by the end of the war in 1945 and Sea Fury orders were reduced to only 100 aircraft, all to be produced by Hawker instead of being shared with Boulton Paul as originally planned. In the meantime the first Sea Fury prototype (SR661) flew on 21 February 1945 powered by a Centaurus XII driving a four-bladed propeller. This aircraft was navalised to the extent that it was fitted with an arrester hook to allow deck-landing trials to take place but it had non-folding wings. Two further prototypes, SR666 and VB857, flew on 12 October 1945 and 31 January 1946. Both of the latter were much more representative of what was to be the production standard with folding wings (although initially manually operated) and five-blade Rotol propellers. Inevitably flight testing revealed a number of problems, the most serious of which were a series of engine failures caused by lubrication difficulties and it was only due to the skill of the Hawker test pilots that none of the prototypes were actually lost. The necessary modifications were eventually applied to a new version of the engine, the Centaurus XVIII, which was to power all production aircraft. There was also a serious problem with the rudder controls. This was eventually cured by a combination of a locking tailwheel, adjusted spring tab settings, and explicit instructions in the pilot's handling notes. Finally, the prototypes and early production aircraft experienced problems during deck-landing trials with the arrester hook and a lengthened and stronger hook eventually provided the solution.

The first of fifty production Sea Fury F.Mk.10s flew in September 1946 and was subsequently employed in deck-landing trials aboard HMS *Victorious*. Other aircraft went to various trials and test units and several were employed by the

A&AEE at Boscombe Down for the clearance weapons and external stores. These included bombs of up to 1,000 lb (one of which could be carried under each wing), napalm tanks, smoke floats, depth charges and mines. Tests were also carried out with underwing rocket launchers and eventually a maximum of sixteen 3 inch rockets with 60 lb warheads could be carried on eight duplex zero-length launchers. The fixed armament comprised four wing-mounted 20 mm Hispano Mk.5 cannon. Finally, instead of ordnance, 45- or 90-gallon drop tanks could be fitted to the underwing attachment points.

Sea Fury F.Mk.10s did reach some frontline units where they were used mainly for training and familiarisation pending the introduction of the definitive production version, the FB.Mk.11. Early production F.Mk.10s had featured a four-bladed propeller and few were fitted with powered wing folding. In the FB.Mk.11 both hydraulic wing folding and a five-bladed propeller were standard and the Fighter Bomber (FB) designation reflected a change in role whereby the Sea Fury was now seen as a multi-role aircraft, considerably enhancing the striking power of the carrier air groups. RATOG had been tested on the earlier aircraft and provision for its use was now standard.

The first operational Sea Fury unit was 807 Squadron based at Eglinton where FB.Mk.10s briefly replaced Seafires in September 1947. However, by February of the following year the squadron began receiving the newer FB.Mk.11 and later embarked in HMS *Theseus* for a while before deploying to the Far East and later becoming heavily involved in Korean War operations. The squadron continued to fly Sea Furies until 1954. A similar story could be told of other Royal Navy frontline squadrons including 802 and 804 Squadrons, which both served in Korea and retained their Sea Furies until March and January 1954 respectively. Other squadrons included 801 Squadron, which operated Sea Furies from March 1951 until January 1955, including a period aboard HMS *Glory* off Korea. Unusually, the Sea Furies had replaced de Havilland Hornet twin-engined fighters. One of the last frontline units to receive Sea Furies was 810 Squadron, which reformed in 1954 and subsequently embarked in one of the new light fleet carriers, HMS *Centaur*, before disbanding after twelve months in March 1955.

The Sea Fury also served with several second line and training units, including 736, 738, 759, 764 and 778 Squadrons, and many of these also flew a two-seat Sea Fury T.Mk.20. This version had been developed by Hawker for the export market but aroused Admiralty interest and sixty were eventually ordered and delivered. The most obvious feature of this version was a second bubble canopy for the instructor who was provided with a prominent periscope sight so that he could monitor the students' gunnery exercises. For this purpose only two 20 mm cannon were fitted. In most respects, however, the T.Mk.20 was similar to the operational versions, although it was not intended for carrier use and therefore was not fitted with an arrester hook. As well as regular Royal Navy squadrons, the Sea Fury also served in significant numbers with some of the RNVR squadrons, including 1831 at Stretton, 1832 at Culham and Benson, 1833 at Bramcote, 1834 at Benson, and 1835 and 1836 also at Culham and Benson. In most cases these retained their Sea Furies until 1955.

The widespread adoption of the Sea Fury by the Royal Navy inevitably led to it equipping the Commonwealth navies that had decided to operate aircraft carriers in the post-war period – Australia and Canada. In the case of the RAN, two Sea Fury squadrons were formed in England in August 1948 and April 1950, these being allocated numbers in the Royal Navy squadron sequence, 805 and 808. Both of these units served in Korea aboard the light fleet carrier HMAS *Sydney*, and also in the British carrier HMS *Venerable*. A third RAN squadron, 850, was briefly formed for Korean operations in January 1953 but was disbanded in August 1954, shortly followed by 808 Squadron the following October. 805 Squadron soldiered on with its Sea Furies until 1958 and even after that a few aircraft were retained for training and target towing until as late as 1962.

The Royal Canadian Navy had planned to form a two-carrier force for Pacific operations but the war ended before these plans could be implemented. However, the light fleet carrier HMS *Warrior* was loaned for a period and 803 Squadron was transferred to the RCN, initially with Seafires and then with Sea Furies in August 1947. Subsequently this squadron served aboard HMCS *Magnificent* and operated Sea Furies until June

1954. In 1951 the sqadron was redesignated VF-870, in conformity with the US Navy system. A sister squadron, 883, was also formed with Seafires and re-equipped with Sea Furies in September 1948, becoming VF-871 in 1951 and retaining its aircraft until as late as August 1956.

It was not only the Commonwealth navies who evinced an interest in the Sea Fury. The Royal Netherlands Navy had placed orders as early as October 1946 for ten F.Mk.50s, which were basically identical to Royal Navy F.Mk.10s. The Dutch aircraft were delivered in 1948/9 and were followed by a further twelve FB.Mk.51s, which corresponded to the British FB.Mk.11. Finally a further twenty-five FB.51s were produced under a licence agreement by Fokker. The early examples were allocated to the GVO (Gevechtsvliegopleiding), the fighter pilot combat school based at Valkenburg. As further aircraft became available two shore-based squadrons, No. 1 and No. 3, were formed, together with 860 Squadron, which subsequently served aboard the carrier HMNS *Karel Doorman*. This was an ex-British Colossus class light fleet carrier that had been purchased in 1948 and replaced the ex-British escort carrier HMS *Nairana*, which had also been named *Karel Doorman* in Dutch service. The Dutch Navy had in fact planned to operate two carriers, which accounted for the relatively large Sea Fury orders, but the funds never became available. The second HMNS *Karel Doorman* had a long career with the RNN and was modernised in 1955–8, subsequently being sold to Argentina in 1968. Under its new name *Veinticinco de Mayo*, she was still in service at the time of the Falklands War and was one of the ships threatening the Royal Navy task force, although in fact she did not come into action and her squadrons eventually operated from shore bases due to problems with the catapults. The Sea Fury remained in service with the RNN until the late 1950s, operating on many occasions from Royal Navy carriers alongside their British counterparts.

Apart from the Netherlands, the Sea Fury had considerable success in the export market with several shore-based air forces, including Iraq and Pakistan, but this aspect of the aircraft is not relevant to this book. Even so, no fewer than 615 Sea Fury FB.Mk.11s were built to Royal Navy contracts (in addition to the fifty F.Mk.10s) but these figures included around 100 supplied to the Australian Navy and another seventy-four to the Canadian Navy. Undoubtedly the Sea Fury was an outstanding success in its role as a naval fighter and represented the pinnacle of piston-engined development before ultimately being superseded by the new generation of jet aircraft. Even then it proved a match in some circumstances, one famously shooting down a MiG-15 over Korea in August 1952. Much of its success was due to the soundness of the basic Hawker design with its strong airframe, rugged wide-track undercarriage, powerful radial engine and superb manoeuvrability and handling. During the intensive Korean War operations the Sea Fury maintained a very high serviceability rate and losses due to deck-landing incidents in difficult conditions were relatively few. Certainly, the aircraft was much better able to put up with the stresses of day-to-day carrier operations than the thoroughbred Seafire that it largely replaced.

Despite its excellent performance, the Sea Fury was not quite the fastest piston-engined fighter to serve with the Royal Navy. That honour went to the graceful de Havilland Sea Hornet F.Mk.20, which, powered by twin Rolls-Royce Merlins, could achieve a maximum speed of 467 mph at an altitude of 22,000 feet, just in excess of the Sea Fury's 460 mph and Seafire FR.47's 452 mph. The original de Havilland DH.103 Hornet was developed for the RAF and was intended for long-range operations in the Pacific. Full-scale development commenced in mid 1943 with the prototype flying in July 1944. De Havilland had always considered the possibility of a naval version and this was one reason why opposite-handed engines and high drag flaps were incorporated from the start, while the wing design had provision for wing folding. Naval interest resulted in Specification N.5/44 and three early production Hornets were taken in hand for modification to naval standards by Heston Aircraft Co. Ltd. The first of these (PX212) flew in April 1945 but, like the second prototype, did not feature the Lockheed hydraulic wing-folding mechanism that was finally incorporated in the third prototype (PX219). This latter aircraft was immediately employed in deck landing trials commencing in August 1945, these proving very successful despite it being the Royal Navy's first carrier-capable single-seat twin piston-engined aircraft. The most significant problem to arise was a weakness with the undercarriage torque links when stressed by side loads, and ultimately the undercarriage legs were redesigned. The handed inward-rotating propellers completely eliminated torque-induced handling problems (although a single-engined landing was a different affair!) while the twin-engined layout with the pilot seated well forward in the slim fuselage gave an unprecedented view of the deck on approach.

Despite the end of the Pacific War in September 1945, a production order was placed and eventually some seventy-eight Sea Hornet F.Mk.20s were produced, the type entering service in late 1946 with 703 Trials Squadron at Lee-on-Solent. The first operational squadron was 801, which reformed at Ford in June 1947 and subsequently served aboard HMS *Implacable* and HMS *Indomitable* with the Home Fleet. Considering the pace of some post-war development programmes, the fact that the Sea Hornet was in service less than two years after the flight of the first prototype is an indication of the priority accorded to this aircraft. The Royal Navy now had a very potent long-range strike fighter, which was armed with four 20 mm cannon and could carry two 1,000 lb bombs. Apart from the basic F.Mk.20, a further forty-three Sea Hornet PR.22 photographic reconnaissance versions were ordered, these being basically similar to the fighter version but with the cannon armament removed and three cameras mounted in the lower fuselage, although the wing hardpoints were retained and could be used for either bombs or drop tanks. With a slightly lighter all-up weight, the PR.22 was no faster but had a better rate of climb, higher service ceiling (37,000 feet) and a longer range (2,050 miles).

Although the Sea Hornet was a great pilot's aeroplane, its size and complexity meant that it was unsuitable for deployment aboard the light fleet carriers that provided the backbone of the Royal Navy's carrier force in the decade after World War II. In addition, the single-engined Sea Fury was capable of carrying out most of the same mission profiles and consequently 801 Squadron's Sea Hornets were retired in March 1951. Re-equipped with Sea Furies, the squadron deployed to Korea the following year. Nevertheless, the graceful twin-engined fighter remained active with Fleet Requirements Units (728 at Malta and 771 at Hurn), as well as training squadrons at Culdrose. The last unit to fly the single-seat Sea Hornet was 728 Squadron at Malta, which used them as late as 1956.

However, another version of the Sea Hornet had a slightly longer operational life, this being the two-seater NF.Mk.21 night fighter. The single-seat fighters such as the Seafire and the later Sea Fury were not radar-equipped and could only act as day interceptors but wartime experience showed the need for a radar-equipped night fighter and this requirement had been partially met by a night fighter version of the single-engined Firefly (described later). However, the concept of a two-seat radar-equipped version of the Sea Hornet held promise of better performance and accordingly Specification N.21/45 was issued. In practice the necessary modifications were not easy to incorporate in the very slim fuselage of the original single-seat fighter, which had been designed to have the minimum cross-section area. The fitting of the radar was not too much of a problem, the ASH radar scanner being contained in a thimble radome at the front of an extended nose fairing, which rather spoilt the aircraft's good looks. The radar navigator was housed in a cramped rearward-facing compartment level with the wing trailing edge and was provided with a small one-piece blister canopy. The tailplane span was increased and, in view of the night fighter role, flame dampers were fitted over the exhaust manifolds. A secondary role of this version was to act as a leader of a strike formation of single-seaters with the navigator responsible for guiding the force to its target and then back to the mother ship.

All these modifications, together with the installation of an arrester hook and other naval equipment, were carried out on the prototype NF.Mk.21 (PX230), which was an RAF Hornet F.Mk.1 converted by Heston Aircraft. This aircraft, which flew on 9 July 1949, did not have folding wings but this was incorporated in the second aircraft (PX239). PX239 also had an extended dorsal fin fillet, a feature that was retrospectively applied to all versions of the Hornet, both RAF and Royal Navy. Surprisingly the modifications and extra equipment had little effect on performance and the NF.Mk.21 was only 5 mph slower than the single-seaters while the quoted rate of climb and service ceiling were both improved, possibly due to the fitting of different Merlin engines (133/134). Carrier trials, including night operations, were successfully carried out aboard HMS *Illustrious* in late 1948 by which time some seventy-eight NF.Mk.21s had been ordered and subsequently 809 Squadron was reformed at Culdrose with Sea Hornet night fighters. Despite being the most complex carrier-based aircraft in service, the squadron's first sea deployment was aboard the light fleet carrier HMS *Vengeance* for short periods in 1950 and 1951. This was mainly because the larger carriers were usually retained in home waters and were often utilised for training, trials and testing. It was not until the first of the new fleet carriers, HMS *Eagle*, commissioned in 1952 that the squadron found a more permanent home. However, its remaining career was relatively short and in May 1954 it was disbanded. Other NF.Mk.21s had been allocated to training squadrons (including 728, 762, 771, 787 and 792) and after its withdrawal from frontline service several examples remained in use for radar observer training and fleet requirements until around 1956. Unfortunately no examples of either the single- or two-seat Sea Hornet were preserved and its absence is noticeable in the otherwise comprehensive collection of aircraft from this period at the Fleet Air Arm Museum at RNAS Yeovilton.

As already mentioned the backbone of the naval fighter force from 1948 onwards until the coming of the jets was the Hawker Sea Fury. Its constant stablemate throughout this period was the Fairey Firefly two-seat single-engined fighter reconnaissance aircraft. The Firefly was originally designed to meet Specification N.5/40, which required a high-performance multi-role aircraft capable of acting as a fighter when required. The prototype flew in December 1941 powered by a 1,730 hp Rolls-Royce Griffon and flight testing quickly confirm the basic soundness of the design and it proved to have the necessary good low-speed handling characteristics vital for safe carrier operation. This was partly due to the adoption of area-increasing Youngman flaps, which could be fully retracted when not required. It was also found that with these flaps in the take-off position, dogfighting manoeuvrability was much improved to the extent that it could out turn a Spitfire at low speeds. The Firefly FR.1 began to equip frontline squadrons from late 1943 onwards and made its operational debut in the course of a series of attacks on the German battleship *Tirpitz* in July 1944. Subsequently three squadrons served with the British Pacific Fleet in action against Japanese forces in the closing stage of the war. Also with the BPF was one squadron equipped with Firefly NF.1 night fighters. The original night fighter variant was the NF.2, which was extensively modified to carry an AI.Mk.X radar. However, only a few were delivered before it was realised that the radar could be mounted in a pod under the standard Firefly, resulting in the NF.1 variant. In all over 850 Mk.1 Fireflies were delivered, including some trainer versions and thirty for the Netherlands Navy.

Production of the FR.1 continued until 1946. Although the Far East squadrons were soon disbanded, at home many of the Barracuda strike squadrons were re-equipped with the Firefly, which, for all practical purposes, was now regarded as the Royal Navy's standard strike aircraft. However, the Mk.1 Firefly was soon to be displaced itself by a new and much improved version of the Firefly, the FR.4. This had a more powerful Griffon 74 engine (prototypes had a Griffon 72), which featured a two-stage supercharger and gave an increase of almost 500 hp, a four-bladed Rotol propeller being fitted to absorb this extra power. There was also a significant number of aerodynamic changes of which the most obvious was the removal of the chin-mounted cooling radiators to forward extensions of the wing centre section so that the nose profile became much more stream-lined. The wingtips were clipped, reducing the wing span by almost 12 feet and improving the rate of roll, and the leading edge of the fin was extended to increase stability. Finally, all aircraft were equipped with two underwing pods, of which the port one provided additional fuel capacity while the starboard one contained the radar antenna, which was previously an optional fitting to be carried in an awkward under fuselage pod. The net result of all these changes was a sparkling increase in performance with the maximum speed rising from 316 mph to 376 mph, although other parameters were very similar.

Development of this version was carried out by four converted Mk.1s starting in mid 1944 but the first production FR.4 did not fly until May 1945. With the run down of squadrons after the War it was not until the autumn of 1947 that the first operational units were formed (Nos 810 and 825 Squadrons) at Eglinton, Northern Ireland. Some 160 FR.4s were delivered, the last on 9 February 1948. Of these forty were destined for the Netherlands Navy where they equipped no fewer than five squadrons at various times, including Nos 2 and 4, which were embarked on the *Karel Doorman* for several deployments. However, the most important post-war Firefly variant was the Mk.5. This variant was produced in several specialised versions based on the standard FR.5, which filled the day fighter reconnaissance role and also acted as a light strike aircraft, while the NF.5 night fighter carried an air-to-air radar in the starboard underwing pod. The third version was the AS.5, which was optimised for the anti-submarine search role for which the Firefly was increasingly used in the later stages of its operational career. A total of 325 Mk.5s were built from December 1947 onwards, the last being delivered on 19 May 1950, and fourteen of these were for the Netherlands Navy. Although there was little external difference between the Mk.4 and 5, the latter introduced power wing folding, a modification introduced in January 1949 and subsequently extended to all Mk.5s.

By this time the Firefly had evolved into a very competent multi-role naval aircraft well suited to carrier operations by virtue of its excellent handling characteristics. Approaches to land on benefited from the Fairey-patented high lift Youngman flaps, which retracted completely into the wing when not in use. During the Korean War this was clearly illustrated by some extremely creditable performances by the squadrons embarked on the light fleet carriers. For example, during her initial deployment, Fireflies aboard HMS *Theseus* made over 1,300 deck landings without accident while maintaining an exceptionally high serviceability rate. Altogether five Royal Navy Firefly squadrons (810, 812, 821, 825, 827) participated in operations off Korea, together with one Australian squadron (817). The last Fireflies to see action were those of 825 Squadron, flying off HMS

Warrior, which supported ground operations against communist forces in Malaya in 1954.

By that time virtually all Royal Navy Mk.5s had been replaced by a new version, the Firefly AS.6, which was a specialised anti-submarine aircraft in which the gun armament was removed, although underwing loads could include sonobuoys and depth charges. The prototype flew on 23 March 1949 and this version entered service with 814 Squadron at RNAS Yeovilton in January 1951. A total of 133 AS.6s were built, the last being delivered in September 1951, and they became the Fleet Air Arm's main anti-submarine aircraft until the turboprop Gannet (also produced by Fairey) became available in 1955. By the time of the Coronation Review in 1953, no fewer than five squadrons flew the Firefly AS.6 as well as several RNVR units. The anti-submarine role grew increasingly important during the early days of the Cold War and this resulted in a further specialised version of the Firefly, the AS.Mk.7, which is described later. Rather ignominiously the last stage of the career of the Firefly, an aircraft which had seen action in World War II and had subsequently formed the backbone of the Royal Navy's post-war carrier groups, was as an unmanned target aircraft. Forty of these were converted Mk.5s, which were designated Firefly U.9, while the U.8 was a modification of the AS.7 of which thirty-four were new build. The last of these was delivered on 26 March 1956, being the 1,702nd and very last Firefly to be built.

When the FR.4 entered service it offered an advance in performance on the wartime Mk.1s and its armament of four 20 mm cannon together with underwing ordnance of bombs or rockets up to a total load of 2,000 lb made it a very respectable strike aircraft. However, the one role that it could not carry out was that of torpedo bomber. During World War II this function had been carried out by the ungainly Fairey Barracuda while units of the British Pacific Fleet were mostly equipped with the American Grumman Avenger. A requirement for a high-performance torpedo bomber was clearly foreseen during the War and this was eventually met by the big and powerful Blackburn Firebrand TF.Mks 4 and 5, which began to enter frontline service in September 1945, too late to see action. At the time it was unique in Royal Navy service in that it was the first single-seat torpedo bomber since the Blackburn Dart biplane of the 1920s. Its other claim to distinction was much more dubious and that was the inordinate time spent in development before it finally entered service. Indeed, the Firebrand had a distinctly chequered history, having been first designed as a high-performance naval fighter to meet specifications drawn up in 1939 and 1940 calling for a replacement for the Gladiators, Rocs and Fulmars then in service. The prototype Firebrand F.Mk.1 flew at Brough on 27 February 1942 and was powered by a 2,035 hp Napier Sabre in-line liquid-cooled engine. Subsequent prototypes carried out deck-landing and armament trials in 1943 but numerous handling problems were highlighted and, in any case, the Ministry of Supply decided that all Sabre engines should be reserved for production of Hawker Typhoons for the RAF. Naval fighter requirements were deemed to be met by the Seafire Mk.III then entering service.

However, the Blackburn aircraft had demonstrated con-siderable potential and was capable of carrying heavy loads. Consequently the second prototype Firebrand, which had been damaged in an accident, was rebuilt as a TF.Mk.II with a widened and strengthened centre section to allow the carriage of a standard 1,850 lb 18 inch airborne torpedo. This flew in March 1943 but only twelve were eventually produced due to restrictions on the supply of Sabre engines and after trials these were used by 708 Squadron at Lee-on-Solent for training and familiarisation. A formal requirement for a high-performance torpedo bomber was crystallised as Specification S.8/43 and to meet this Blackburn produced the Firebrand TF.Mk.III, in which the Sabre engine was replaced by a 2,400 hp Bristol Centaurus VII eighteen-cylinder twin-row air-cooled radial engine. The prototype flew on 21 December 1944 but only twenty-four were produced as this version once again displayed some undesirable handling characteristics, of which the most noticeable was poor directional stability on take-off, caused by the massive torque of the powerful Centaurus engine.

The Firebrand's handling and performance was finally brought up to acceptable standards in the TF.Mk.4, which first flew on 17 May 1945, and 104 of this variant were eventually delivered. The TF.Mk.4 had a much larger fin, which was offset by 3 degrees in order to overcome the directional stability problems. It also introduced other changes that considerably enhanced the aircraft in its operational role. These included a teardrop clear-view canopy, retractable wing spoilers and provision for a wide range of ordnance. As an alternative to the torpedo, two 2000 lb bombs or sixteen 60 lb rocket projectiles could be carried as well as 45-gallon underwing or 100-gallon centreline drop tanks. The torpedo itself was carried on a unique two-position crutch in which the weapon was tilted down for take-off in order to prevent its tail dragging on the ground, but was tilted up parallel to the fuselage when the undercarriage was retracted.

The final Firebrand production version was the TF.Mk.5, which first flew in 1945 and entered service with 813 Squadron in April 1947. Sixty-eight were built and some forty Mk.4s were also converted to Mk.5 standard. The Mk.5 differed from the earlier version in detail only, the changes being aimed at further improving handling characteristics and included horn balanced ailerons and increased-span aileron tabs. Finally, the TF.Mk.5A was introduced, which featured hydraulically powered ailerons, one of the first application of this type of control system in a British frontline aircraft. Although this was a great improvement, the Firebrand was regarded as a difficult aircraft to fly and, in general, only experienced pilots were drafted to the two operational squadrons, 813 and 827. It was not only the handling characteristics that made carrier approaches and landings difficult. The problem was compounded by the fact that the pilot was positioned some 18 feet behind the big radial engine and his forward view in the landing attitude was virtually nil! The fact that the Firebrand was known among its pilots as the 'Firebrick' is perhaps all that needs to be said about their opinion of the aircraft. These features, together with the sheer size of the aircraft, made the Firebrand unsuitable for use aboard the light fleet carriers and 813 Squadron confined its operations to the large fleet carriers (*Illustrious*, *Implacable* and *Indefatigable*), mostly in home waters. 827 Squadron was reformed with Firebrands in 1952 specifically for service aboard the new carrier HMS *Eagle*, which entered service that year. In 1953 the

Firebrand was withdrawn from frontline service when it was replaced by the turboprop Wyvern, an aircraft whose gestation was even more convoluted and extended than that of the Firebrand, as will be described later.

Early in the Firebrand's development some of the aircraft's drawbacks were apparent and consequently Specification S.28/43 was issued in February 1944, which asked for an improved view for the pilot and a new wing. Blackburn responded with its B-48 project, which was based on the Firebrand and subsequently used the name Firecrest, although this was never adopted officially. (Post-war, the designation Y.A.1 was also applied under a new system brought in by the SBAC.) The problem of the pilot's view was solved by raising the cockpit and moving it forward, exchanging position with some of the fuselage fuel tanks. The new wing was substantially different and was of thin laminar flow profile. In plan view the leading edge was sharply tapered while the trailing edge was almost straight. The centre section had a pronounced anhedral, while outboard of the wheel bays the wing had nine degrees of anhedral, the whole similar in concept to the American F4U Corsair, although not quite so pronounced. However, the reasoning was the same, the gull wing layout permitting a shorter undercarriage while still leaving ground clearance for the five-bladed Rotol propeller. Another change from the Firebrand was the wing-folding arrangement. Whereas the earlier aircraft's wings were twisted to fold alongside the fuselage, in the Firecrest they folded upwards just outboard of the wheels and the outer section and ailerons then folded downwards and inwards – a unique arrangement not repeated on any other naval aircraft. Fowler high-lift flaps were fitted on the trailing edge and electrically operated trim tabs were fitted on all moving control surfaces. Hydraulically actuated dive brakes were also fitted above and below the wing.

The first Firecrest did not fly until well after the end of the war, in fact as late as 1 April 1947 and a second aircraft was only partially completed. The aircraft was slightly lighter than the Firebrand and with essentially the same engine (Centaurus 59) there was a modest increase in performance. The improved pilot's position attracted favourable comments. However, the handling characteristics were little better. The third prototype featured the powered ailerons then being introduced on the Firebrand, and in connection with this the dihedral of the outer wings was reduced to three degrees. This aircraft flew in 1948 but overall there was little improvement. Nevertheless continued test flying yielded much useful information on the operation of powered flying controls, some of which was applied to the Firebrand TF.5A.

As a military aircraft, the Firecrest did not have any fixed armament but provision was made for two 0.5 inch machine-guns to be carried in pods attached to underwing hard points. Alternative loads included 500 lb bombs or rocket projectiles while a torpedo could be carried under the centre fuselage. However, the aircraft offered little improvement over the Firebrand and by the late 1940s the future of combat aircraft undoubtedly lay with the new jets and so no production con-tracts were forthcoming. Blackburn did some further design work on Firecrest versions that would be powered by the new turboprops such as the Armstrong Siddeley Python or the Rolls-Royce Clyde driving six-bladed contra rotating propellers. This work led eventually to the Blackburn B-54, originally envisaged as a strike aircraft but which eventually emerged as an anti-submarine patrol aircraft.

While Blackburn persevered with the Firebrand and Fire-crest, the Fairey company also built and flew a heavy strike bomber in the form of the Fairey Spearfish. This was intended as a successor to the Barracuda and was designed to meet Specification O.5/43 for a two-seat multi-role torpedo/dive bomber. The initial contract was placed in August 1943 and the first prototype was flown on 5 July 1945. The Spearfish was a very clean and functional-looking aircraft powered by a single 2,600 hp Bristol Centaurus radial engine with the two crew members accommodated under a clear-view canopy set in line with the leading edge of the wing. For the first time on a British naval aircraft, the offensive load was all carried in an internal bomb bay. This could consist of either the standard British 18 inch, or the larger American 22.4 inch, torpedo. Alternatively four 500 lb, or one 2,000 lb bomb, or four depth charges could be carried, while another possibility was a 180-gallon long-range fuel tank for ferry purposes. Two wing-mounted 0.5 inch machine-guns were carried while another two were intended to be carried in a Frazer Nash 95 remote-controlled dorsal barbette to be installed immediately behind the cockpit canopy. Sixteen 3 inch, 60 lb rockets could be carried in zero-length underwing launchers (the only armament carried externally). Provision was made for a Mk.15 ASV radar to be mounted in a retractable 'dustbin' fairing immediately behind the weapons bay, which occupied most of the lower fuselage. To leave space for this arrangement, a mid-mounted wing configuration was adopted, which led to a longer than usual undercarriage that retracted outwards. Hydraulic wing folding was standard.

In terms of performance the Spearfish offered little improvement over its predecessors but it showed every promise of having better handling characteristics and being more adaptable to a variety of roles, and initially production orders were placed for up to 152 aircraft. However, these were subsequently cancelled after the end of the War and only three prototypes and one production aircraft were flown (the latter fitted with the Frazer Nash barbette), the first two in 1945 but the last two not until 1947. These were used for various tests and operational trials, at least one still flying as late as 1952 with the carrier trials unit at Lee-on-Solent and Ford.

As already described, the Blackburn Firebrand was the Royal Navy's main torpedo strike aircraft throughout most of the period covered by this book. However, in 1953 its replacement finally entered service after no fewer than seven years' testing and development following its first flight in December 1946. In addition this aircraft represented a bridge between the old and the new as, although propeller-driven, its powerplant was a gas turbine. This was the Westland Wyvern, which had originally been conceived in 1944 as a high-performance naval fighter designed around the 3,500 hp twenty-four-cylinder Rolls-Royce Eagle piston engine, although the potential to install one of the new turboprop engines was not ignored. Naval Specification N.11/44 was issued and this called for a maximum speed in excess of 500 mph and the ability to carry a wide range of weapons, including one 2,000 or three 1,000 lb bombs, or a single Mk.XVIII torpedo. This was an exacting requirement and even as the TF.Mk.1 prototypes were under construction, a decision was made to produce a Mk.2 version powered by the new Rolls-Royce Clyde turboprop. Almost immediately the programme was affected by severe delays of both the Eagle and Clyde engines but the piston-engined prototype finally flew on 16 December 1946 and subsequently a total of fifteen TF.Mk.1s were produced. However, the Eagle engine and its associated twin four-bladed contra-rotating propellers suffered numerous problems and the prototype was lost in October 1947 in an accident caused by propeller failure, which also killed its pilot. By this time priority was being given to the turboprop-powered TF.Mk.2 but now the Armstrong Siddeley Python was the favoured powerplant, its development being ahead of the Rolls-Royce Clyde. One Clyde-engined TF.Mk.2 was flown on 18 January 1949, followed by a Python-engined example on 22 March 1949. On paper the Clyde was the more powerful with 4,000-shp being absorbed by the propellers and the gas turbine providing an additional 1,550 lb residual thrust, while the equivalent figures for the Python were 3,560 shp and 1,100 lb thrust. The Python, however, was comparatively more mature and as the Clyde's problems became apparent it was thought to be the better option, although its installation in the Wyvern required alterations to the fuselage design including a 21 inch increase in length.

A batch of twenty Python-powered Wyverns were ordered and the first of these flew on 16 February 1950. Almost immediately numerous problems were manifest and a number of changes were made to cope with various handling problems. The most noticeable of these was a substantial increase in the height of the tail fin and the cutting back of the lip of the annular air intake exposing the central fairing covering the gearbox. However, the most serious problem and one that almost brought the programme to a complete halt was an uncontrollable tendency for the engine and propeller combination to surge. This took months of trials and testing before a solution was found comprising a combination of modifications to the engine by Armstrong Siddeley and a new constant speed propeller control unit by the manufacturers, Rotol. The problem was not resolved until April 1952 by which time deliveries of the definitive production version, the Wyvern S.4, had begun. In fact, the last seven TF.2s were completed as S.4s and a further ninety were delivered as new-build aircraft. Even then, all was not well and final tests at Boscombe Down revealed rudder control problems, which were eventually cured by the addition of small finlets on the leading edge of the tailplane. The Wyvern S.4 eventually entered service with 813 NAS in May 1953 but engine limitations kept the unit shore-bound until September 1954 when it embarked in HMS *Albion*. Even then there were unforeseen problems, including an alarming tendency for the engine to flame out during the acceleration of catapult launches, resulting in several aircraft being ditched (and also the world's first underwater ejection in which the pilot survived a ditching after being shot off into the sea). The Wyvern's frontline service life was very short (especially in comparison with its lengthy development time), the last operational squadron disbanding in February 1957. Perhaps the highlight of its career was in October and November 1956 when 830 Squadron embarked in HMS *Eagle* carried out seventy-nine sorties for the loss of two aircraft during the ill-fated Suez operations.

Despite its numerous problems and short operational career, the Wyvern is significant because it was the Royal Navy's last propeller-driven fighter or strike aircraft and therefore represented the end of an era. During the late 1940s and early 1950s the Sea Fury and the Firefly, backed up by a few Firebrands, had provided much of the Royal Navy's frontline strength. Although aircraft such as the Seafang, Firecrest and Spearfish explored the ultimate performance envelope of the piston-engined propeller-driven combat aircraft, production orders could not be justified as it became obvious that the future lay with the new generation of jet aircraft then under development and coming into service.

2

PROPELLER SWANSONG – US NAVY

By the closing stages of the Pacific War, the US Navy's Essex class carriers, which formed the backbone of the mighty 3rd Fleet, were equipped with either Grumman F6F Hellcat or Vought F4U Corsair fighters, Grumman TBF/TBM torpedo bombers and Curtis SB2C Helldiver dive bombers. The older and slower Grumman F4F Wildcat was still in frontline service aboard many of the smaller escort carriers, and examples of all these aircraft (except the Helldiver) had been supplied in considerable numbers to the Royal Navy.

The tough little Wildcat first flew in September 1937 and was developed into a rugged and reliable naval fighter, which was supplied to the Royal Navy from 1940 onwards (initially known as the Martlet), while the US Navy took delivery of the F4F-3 from December 1940 onwards. This version was powered by a 1,200 hp Pratt & Whitney R-1830-76 Twin Wasp, which gave a top speed of 335 mph and a 3,100 feet/min initial rate of climb. When combined with excellent handling and manoeuvrability, the outcome was to make the Wildcat one of the best naval fighters of the time. By the time of the attack on Pearl Harbor virtually all US Navy fighter squadrons were equipped with the type, which remained the only carrier-borne fighter until the introduction of the Hellcat in early 1943, and it bore the brunt of the air fighting in all the early Pacific battles. A total of 7,825 Wildcats had been built when production ceased in August 1945, although only 1,988 were completed by the parent Grumman company, the remainder being produced as the FM-1 and FM-2 by the Eastern Aircraft division of General Motors. After 1945, the Wildcat was rapidly withdrawn from frontline service and despite its very distinguished war record, it has little relevance to the overall theme of this book.

However, the Wildcat ancestry was vital to the development of its successor, the Grumman F6F Hellcat, which was to become one of the most successful of the Allied fighters during World War II. The origins of the Hellcat went back to 1938 when Grumman designers looked at the possibility of installing a 1,600 hp Wright R-2600 radial engine in the F4F Wildcat. But it soon became apparent that a larger airframe would be required and the project lapsed until 1940 when reports filtering back from the air war in Europe indicated that a replacement for the Wildcat would soon be required. At that time the US Navy was placing its hopes on the new Vought F4U Corsair, which had flown in that year. Consequently, Grumman were left to their own devices, the project being regarded merely as a back-up in case of problems with the Corsair. In fact, as will be related, although the Corsair offered a much better performance its handling characteristics initially caused the Navy to rule out shipboard operations and the new Grumman F6F Hellcat began to assume a much greater importance. This led to orders for 1,080 F6F-1s being placed in January 1942, almost six months before the prototype XF6F-1 made its first flight on 26 June 1942. This was powered by the Wright R-2600 radial engine rated at 1,700 hp for take-off. A second prototype, designated XF6F-2, was to have been fitted with a turbosupercharged XR-2600-10 but instead was powered by a 2,000 hp Pratt & Whitney R-2800-10 Double Wasp when it flew on 30 July 1942 (the same engine also powered the Corsair). In this guise it was designated XF6F-3 and formed the basis for the production F6F-3 of which no fewer than 4,402 were built, the last being delivered in April 1943.

The Hellcat inherited a reputation for strength and reliability from the Wildcat and the early production aircraft were allocated to the new air groups being formed in early 1943 for service aboard the Essex class carriers that were then coming into service. The first US Navy squadron to equip with Hellcats was VF-9 at NAS Oceana, Virginia, and it was destined for service aboard the USS *Essex*. By August 1943 the Hellcat was in action for the first time when ships of Task Force 15 including

Essex, *Yorktown* (VF-5 embarked), and *Independence* (VF-6 and VF-22 embarked) attacked Marcus island. The Hellcat quickly established its superiority over the hitherto much feared Mitsubishi Zero. As the campaign across the Pacific gathered pace Hellcats became the main US Navy shipboard fighter and in Operation *Forager*, the occupation of the Mariana Islands, no fewer than fifteen carriers of Task Force 58 carried a total of 896 aircraft of which 467 were Hellcats. This figure included two dozen F6F-3Ns, which were modified to act as night fighters with an AN/APS-6 radar set in a radome fitted below the leading edge of the starboard wing, and some 229 examples of this version were produced by modifying standard -3s.

The F6F-3 had a maximum speed of 376 mph at 17,300 feet and a range of 1,090 miles on standard tankage, which could be extended to 1,590 miles by means of an under fuselage 125-gallon drop tank. The standard armament was six wing-mounted 0.5 inch machine-guns and two 1,000 lb bombs or six 5 inch rockets could be carried on underwing hardpoint – giving the Hellcat a useful secondary role as a fighter-bomber. The next major version was the F6F-5, which replaced the -3 on the production lines from April 1944. No fewer than 7,668 of this variant were built between then and 21 November 1945 when the last Hellcat left the Grumman factory. In order not to disrupt the production programme, the F6F-5 incorporated only detailed improvements over its predecessor, these included a redesigned engine cowling, improved windshield design, spring-tabbed ailerons and strengthened tail surfaces. Armour protection for the pilot was increased but plans to fit the aircraft with a clear-vision bubble canopy and a cut-down rear fuselage were not adopted. A night fighter version, the F6F-5N, was designed, some 1,432 being produced, and in some of these a single 20 mm cannon replaced the two inboard machine-guns in each wing. By early 1945 the F6F-5 had replaced earlier versions aboard all frontline carriers and by the end of the War US Navy and USMC Hellcats, both carrier- and shore-based, had accounted for the destruction of 5,156 enemy aircraft. All of the US Navy's top scoring aces, led by Cdr David McCampbell USN with thirty-four victories, were Hellcat pilots.

Inevitably, with the end of hostilities in 1945, the Navy underwent a reduction in frontline strength and many Hellcat squadrons were disbanded. However, some remained active aboard carriers until 1948 before they were finally replaced by Corsairs and Bearcats, although reserve squadrons retained Hellcats until the end of the Korean War in 1953. One high-profile unit to fly Hellcats in the post-war era was the now famous Blue Angels, the US Navy's aerial demonstration team, which was formed with Hellcats in July 1946. An exception to the general run down of US Navy Hellcat units was the night fighter F6F-5N, which was in service with Composite Squadron VC-4 as late as mid 1954. The availability of great numbers of Hellcat airframes led to them being extensively used as pilotless target drones under the designation F6F-3K or -5K and some of these were in use up to around 1959 – a rather sad end for one of the most significant American aircraft of the World War II era. In passing, it is worth noting that 1,182 Hellcats were transferred to the Royal Navy. They first entered service in July 1943 with 800 Squadron, which was later embarked on the escort carrier HMS *Emperor*, and carried out shipping strikes of the Norwegian coast towards the end of that year. Subsequently most Royal Navy Hellcats served with units of the Eastern and Pacific Fleets.

One version of the Hellcat that did not see service was the projected F6F-4, which was a proposed lightened aircraft with reduced armament and fuel tankage and was intended for use aboard the smaller escort carriers. In the event the FM-2 Wildcat met this requirement and the F6F-4 did not fly. However, the idea of a lightweight high-performance fighter capable of operating from both large and small carriers was resurrected by Roy Grumman himself in July 1943 when he set out a specification for a Wildcat-sized aircraft with a supercharged R-2800 engine, a gross weight of 8,500 lb and a clear-view bubble canopy. A wide-track undercarriage would make for ease of deck landing and provide a long-enough oleo leg to allow adequate clearance for the large four-bladed propeller that would be fitted. Under the designation Design 58, design work began straight away and details were passed to the Navy's Bureau of Aeronautics who evinced an immediate interest. In fact, the proposal had come at an interesting time as the US Navy had already ordered its first all-jet fighter (McDonnell XFD-1) but there were obviously serious reservations about the untried concept of jets at sea and the idea of a lightweight fast-climbing piston-engined interceptor proved an ideal back-up. In this context the US Navy was not particularly interested in the ability to operate from escort carriers, and intended the new aircraft (now designated XF8F-1 Bearcat) to operate from the new Midway class carriers then under construction.

The first of two Bearcat prototypes had its maiden flight on 31 August 1944, less than a year after the initial concept and only eight months after the initial contract was signed in November 1943. Initial flight tests showed some directional stability problems, which resulted in the addition of a dorsal fin and increased tailplane span on subsequent production aircraft. Overall, however, the Bearcat was an outstandingly successful design with an initial rate of climb in excess of 4,500 feet/min, substantially in excess of figures achieved by either the Hellcat or Corsair. Production F8F-1 Bearcats had a maximum speed of 434 mph at 20,000 feet, again much faster than the Hellcat, although not as quick as the Corsair. The lightweight airframe allowed a substantial underwing ordnance load to be carried, initially two 1,000 lb bombs or two 11.75 inch Tiny Time rockets, while later production aircraft had additional outboard racks for four 100 lb bombs or 5 inch HVAR rockets. A 150-gallon drop tank could be carried under the fuselage centreline. With a smaller wing than the Hellcat, the fixed armament comprised only four 0.5 inch machine-guns, although these could be supplemented by underwing twin 0.5 inch gun pods carried instead of the 1,000 lb bombs.

In a break with Grumman tradition, the outboard wing section folded upwards for carrier stowage instead of being folded back against the fuselage sides. A rather odd feature was that the outer wingtips were designed to break off if the pilot exceeded the airframe g force limitations. Unfortunately, what was intended as a safety feature resulted in the loss of at least one Bearcat when only one wingtip broke off during flight testing. Subsequently, explosive bolts were fitted to ensure that both sections came away, leaving the aircraft in a symmetrical configuration. Eventually the feature was discarded and the wings were permanently strengthened at the break point.

Almost as soon as the prototypes were flying, orders were placed for 2,023 Bearcats in October 1944. A further 1,876 were ordered in February 1945 from General Motors under the designation F3M-1. Inevitably these orders were severely curtailed with the end of the War, the General Motors contract being cancelled and the Grumman order reduced to 770 aircraft. Nevertheless, deliveries began to build up early in 1945 and the first squadron (VF-19) began to equip with Bearcats in May 1945. This squadron was embarked on the light carrier USS *Langley* en route for the Western Pacific when the War ended so that the Bearcat did not see action during World War II. Nevertheless, it rapidly replaced the Hellcat in frontline squadrons and by the end of 1948 no fewer than twenty-four US Navy squadrons were flying Bearcats. These included some F8F-1B cannon-armed aircraft, 124 of which were built in addition to the original Grumman order. During the post-war period the improved F8F-2 was developed and flew on 11 June 1947. This had a slightly more powerful R-2800-34W engine, cannon armament was standard and a noticeably taller tail fin was fitted. Some 293 of this variant were produced between November 1947 and April 1949 and a further sixty F8F-2P photographic reconnaissance Bearcats were also produced, the last of these being delivered in May 1949 when Bearcat production ended. As with the Hellcat, efforts were made to produce a night fighter version of the Bearcat but the wing construction made it unsuitable for mounting a radome faired into the leading edge. The radar was therefore carried in an underwing pod, which created significant drag and reduced performance accordingly. Consequently only twelve Bearcats were modified as F8F-1Ns and a further twelve F8F-2Ns were built, but these saw little service with operational units (VCN-1 and VCN-2) and were mostly relegated to training duties.

In fact, the Bearcat's frontline career with the US Navy was relatively brief and it was never deployed by any Marine Corps squadrons. By the time of the outbreak of the Korean War, it was already being replaced as an interceptor by the first generation of Navy jet fighters. As a ground support and attack aircraft it was outclassed by the larger Corsair, which had much superior payload/range characteristics, and consequently it was not deployed to the Seventh Fleet to operate off Korea. Those squadrons embarked on carriers of the Sixth Fleet in the Mediterranean re-equipped with jets during this period and even US-based Reserve squadrons only flew Bearcats until 1953, the last units being VF-859 and VF-921. Although this was effectively the end of the Bearcat's naval career, it did see considerable action ashore with the French Air Force in Indo-China and also the air force of South Vietnam. In addition the Royal Thai Air Force received around 129 Bearcats, including some ex-French aircraft.

Although the various Grumman piston-engined fighters served the US Navy well in great numbers, the aircraft that outlived them all and was probably the finest carrier-based fighter produced in World War II was the Vought F4U Corsair. By the standards of the day when it first flew in May 1940 it was a large and heavy aircraft. However, it was the first American fighter to be powered by the new Pratt & Whitney R-2800 Double Wasp air-cooled radial engine and this enabled the Corsair to achieve speeds in excess of 400 mph, faster than anything then available to the Air Force. The origins of this remarkable aircraft went back to a naval specification issued in 1938 for a new single-seat high-speed naval fighter with a per-formance equal to land-based contemporaries. The new 1,850 hp R-2800-4 engine offered the necessary power but needed a large diameter propeller. The Corsair's distinctive inverted gull wing configuration was adopted so that the necessary propeller ground clearance could be achieved without unnecessarily long and heavy undercarriage legs. An incidental benefit of this layout was that the wing root joined the fuselage at right angles, which substantially reduced aerodynamic drag at that point. The test flying programme confirmed the Corsair's excellent per-formance, which was considerably superior to the Grumman Wildcat's, then the Navy's standard fighter. However, inform-ation coming out of the air war in Europe pointed to the fact that several significant changes were needed if the Corsair was to be effective in combat. The most pressing was the need to increase the armament which, on the prototype, comprised one

0.3 inch and one 0.5 inch machine-gun mounted in the forward fuselage and a single 0.5 inch gun in each wing, while a novel feature was the provision of compartments in the wing for up to ten small bombs intended for use against bomber formations. This whole arrangement was swept away and in its place was set a battery of six 0.5 inch machine-guns, three in each wing, and all with a greatly increased ammunition supply. However, this all took up space previously occupied by fuel tanks and a new 197-gallon self-sealing fuel tank was set into the fuselage immediately behind the engine. This necessitated moving the cockpit some three feet further back, giving the Corsair its distinctive long-nosed appearance. Other modifications included 155 lb of armour protection, a bullet-proof windscreen and increased aileron span. Naturally, the empty weight rose considerably, from 7,505 lb of the prototype XF4U to 8,982 lb for the eventual production F4U-1. But this was offset by the more powerful R-2800-8 engine rated at 2,000 hp that was fitted, giving a maximum speed of 417 mph at 20,000 feet and an initial rate of climb of 2,890 feet/min.

Such development took time and it was not until 30 June 1941 that an order for 584 F4U-1 Corsairs was placed and it was almost a year later before the first production aircraft were delivered. By the end of 1942 the US Marine Corps squadron VMF-14 was fully operational and the first US Navy squadron VF-12 was working up. However, carrier trials aboard the USS *Sangamon* in September 1942 revealed several problems associated with shipboard operations, the most significant of which was the pilot's poor forward view over the long nose. The aircraft also had a tendency to swing violently on touchdown and the stiff undercarriage made it prone to bouncing so that arrester wires were missed and barrier engagements were not uncommon. Consequently the US Navy decided that the Corsair was not suitable for deck operations and its use was restricted to land-based USMC and US Navy units. Indeed, some Corsairs were actually produced without folding wings. Despite these reservations, the Corsair was otherwise an immediate success as it went into action in the fierce fighting around the Solomon Islands and New Georgia in the Southern Pacific. By August 1943 eight Marine squadrons were flying Corsairs in the theatre and land-based US Navy units were also becoming operational. However, it was not until further carrier trials were carried out aboard the escort carrier USS *Gambier Bay* in April 1944 that the US Navy finally approved the Corsair for carrier deployment and it finally began to join the air groups aboard the Essex class carriers for the major operations against the Philippines and the Japanese home islands of Okinawa and Iwo Jima.

The Corsair was supplied in significant numbers to the Royal Navy whose squadrons began forming in the USA in mid 1943 – by the end of that year no fewer than eight had been formed and ultimately there were nineteen British Corsair squadrons. The Royal Navy was delighted to get such a potent aeroplane especially as its performance was as good as anything the RAF had and, more importantly, it could hold its own against the German Messerschmitts and Focke-Wulfes. Consequently they had no hesitation in getting the aircraft to sea from August 1943 onwards. A modification quickly introduced was the fitting of a bulged clear-view canopy, which allowed the pilot's seat to be raised and much improved visibility, both for carrier operations and combat situations. A modified long stroke undercarriage reduced the tendency to bounce and British Corsairs had the wingspan reduced by 16 inches to permit stowage in the hangars of British carriers. Royal Navy Corsairs were first in action on 3 April 1944 when 1834 Squadron flying from HMS *Victorious* formed part of the escort for the highly successful strike against the battleship *Tirpitz*. Eventually the Royal Navy had nineteen Corsair squadrons and the aircraft was widely deployed with the British Pacific Fleet in the closing stages of the war against Japan. Unfortunately, as these aircraft were supplied to Britain under Lend-Lease, they had to be returned or destroyed at the end of hostilities and consequently played no role in the post-war Royal Navy. As has been recounted, there was a gap of two years before the Hawker Sea Fury, which offered a similar performance, entered frontline service.

The US Navy, of course, had no such problems and the Corsair was to remain in frontline service for almost another decade. During World War II Vought built a total of 4,699 F4U-1s but this figure was boosted by another 738 built by Brewster under the designation FBA-1 and 4,007 by Goodyear (designation FG-1).

As already recorded, 2,012 of these went to the Royal Navy and a further 370 to the RNZAF. Almost half of the overall total were produced as fighter-bombers with provision for an under fuselage drop tank and two 1,000 lb bombs or eight 5 inch rockets on underwing hardpoints. Depending on the manufacturer, these were designated F4U-1D or FG-1D.

The F4U-2 was a night fighter version with a radome on the starboard wingtip and was fitted with an autopilot, while armament was altered to four 20 mm cannon. Some thirty-two examples were produced by the Naval Aircraft Factory by converting F4U-1s. The designation XF4U-3 was applied to three prototypes powered by a turbosupercharged version of the Double Wasp but the first did not fly until 1946 and only a single example of the equivalent Goodyear FG-3 was completed. Consequently the next major production version of the Corsair was the F4U-4, which was powered by a 2,100 hp Pratt & Whitney R-2800-18W. This engine boosted the top speed to 446 mph at 26,000 feet and improved the initial rate of climb to 3,870 feet/min. The prototype flew on 19 April 1944 and it quickly replaced the earlier versions on the production lines. A total of 2,357 were built with production continuing, despite the end of the War, until 1947. Although most of these were standard fighter-bombers, there were 297 F4U-4Bs with four 20 mm cannon and nine F4U-4P photo reconnaissance versions. Only a single F4U-4N night fighter was produced.

Even then, the Corsair had the potential for further development in the post-war era and the first new variant was the F4U-5, which standardised on the four cannon armament and was powered by the 2,300 hp R-2800-32W fitted with a two-stage supercharger. A total of 568 F4U-5s were ordered by the US Navy but significantly over half of these (315) were built as F4U-5N or -5NL night fighters, the rest being standard fighter-bombers except for thirty F4U-5Ps. The more powerful engine naturally increased performance, the top speed rising to 470 mph at 26,800 feet, but also allowed an increase in ordnance payload, which could now consist of up to 1,600 lb of bombs under each wing together with a 2000 lb bomb on the centreline rack in lieu of a drop tank. The F4U-5 entered service from 1947 onwards and subsequently played an important role in the composition of US Navy carrier air groups. The F4U-5N in particular provided the essential night and all-weather capability that the new jets were not able to provide, while the propeller-driven fighter-bomber was cheaper to operate and just as effective in the ground attack role. In fact, a specialised ground attack version, the F4U-6, was produced for the USMC squadrons, which had increased armour protection for the pilot and could carry up to four 1,000 lb bombs. As it would spend most of its operational life at low level, only a single-stage 2,300 hp R-2800-83W was fitted and overall performance was substantially reduced. Some 111 of this variant, designated AU-1 in service, were built and they served extensively with the USMC shore-based units in Korea.

The importance of the Corsair can be gauged by the fact that during the time of the Korean War (1950–53), a total of twenty-eight Navy and seven USMC squadrons flew this aircraft and, perhaps surprisingly, the only official US Navy ace of the war flew the F4U-5N night fighter. This was Lt Guy P Bordelon USN who shot down five Communist aircraft while operating as apart of a detachment of VC-3 aboard the USS *Princeton*. However, once the fighting stopped in July 1953, the operational units quickly returned home and were disbanded. By the end of 1954 only two US Navy squadrons (VC-3 and VC-4) still flew Corsairs, these finally disbanding in the following year.

The final production version of the Corsair was the F4U-7 of which ninety-four were built for the French Navy (*Aeronavale*), these being supplied under MDAP arrangements. Although basically similar to the AU-1, they were powered by the two-stage R-2800-18W, which went some way to restoring the performance at medium and high altitudes. The last of these left the Dallas factory in 1952, making the Corsair the last of the World War II era fighters to remain in production.

As already recounted, substantial Corsair production contracts were awarded to Goodyear during World War II, these being built as the FG-1 and FG-4. In the later stages of the Pacific War an urgent requirement arose for a high-performance low-altitude interceptor to counteract the alarming Japanese kamikaze attacks. With the Vought company fully stretched with development and production of the standard Corsairs,

Goodyear was tasked with the project in early 1944. The engine specified was the 3,000 hp Pratt & Whitney R-4360 air-cooled radial engine. Retaining the same basic layout as the Corsair, the fuselage of the new F2G was extensively redesigned to take the much more powerful engine and featured a clear-view bubble canopy, which had been proposed earlier, but not adopted, for production FG-1A Corsairs. Flight tests revealed a top speed of almost 400 mph at sea level and an initial rate of climb of 4,400 feet/min – roughly comparable to the contemporary Bearcat but the F2G was better suited to the secondary fighter-bomber role. In March 1944 contracts were signed for 418 F2G-1s and ten F2G-2s, the latter being equipped for carrier operations whereas the standard F2G-1 was intended for land-based operations with the USMC. By August 1945 only five of each variant had been completed and with the end of the War the remaining orders were cancelled, as indeed were all outstanding Goodyear contracts for FG-4 Corsairs.

With well proven aircraft such as the Bearcat and Corsair available after 1945, the US Navy had little desire to continue development of other piston-engined fighters now that a new generation of jet fighters was on the horizon. Consequently the Goodyear F2G was not the only design to be sidelined at that time and another to suffer the same fate was the little known Boeing XF8B-1. During and after World War II Boeing made its name as the builder of large bombers and transport aircraft but it should not be forgotten that the company had produced some very successful fighters in the 1920s and 30s. Consequently, when the US Navy issued a requirement for a long-range shipboard fighter in 1943, Boeing responded with a very large and heavy single-seat fighter powered by a single 3,000 hp Pratt & Whitney R-4360-10. Three prototypes were ordered as the XF8B-1 but only one flew before the end of the War (27 November 1944), although the other two were subsequently completed at a later date. With a span of 54 feet and a length of over 43 feet, together with a loaded weight of over 20,000 lb, the XF8B-1 was, as perhaps befitted a Boeing product, the heaviest carrier-based aircraft produced during the War. The R-4360-10 engine was a complex four-row radial driving two three-bladed contra-rotating propellers, the whole installation being designed as a removable 'power egg' for ease of maintenance. The top speed was only 432 mph at 27,000 feet and the initial rate of climb was a sluggish 2,800 feet/min but the aircraft's great attribute was a maximum range of over 3,000 miles. In fact, the XF8B-1 was really a multi-purpose aircraft despite the fighter designation and could carry a 6,400 lb bomb load distributed between an internal bomb bay and underwing hardpoints, or two 2,000 lb torpedoes. Viewed in that light the Boeing design had obvious potential but by 1945 the long-range requirement was not so important as US carriers were now powerful enough to operate close to enemy shores and the future obviously lay with jet aircraft. Consequently no production orders were forthcoming and the prototypes were put into storage, although one was briefly evaluated by the USAAF at Wright Field.

The US Navy did have one other piston-engined fighter in service at the end of 1945. This was the twin-engined Grumman F7F Tigercat, which had flown in prototype form in December 1943 and was initially intended to operate from the decks of the new large Midway class carriers then being laid down. This interesting aircraft was the first twin-engined carrier aircraft to go into production and was also the first to feature a tricycle undercarriage. Powered by a pair of 2,100 hp Pratt & Whitney R-2800-22W radial engines, the F7F-1 was capable of over 400 mph and was heavily armed with a battery of four 20 mm cannon in the wing roots and four 0.5 inch machine-guns in the nose. In addition, it could carry a standard torpedo under the fuselage or two 1,000 lb bombs on underwing racks. It was ordered into production for the USMC and deliveries began in April 1944, although due to various factors the Tigercat never saw operational service in World War II. Only thirty-four single-seat F7F-1s were delivered before production changed to the F7F-2N, a two-seat night fighter with AI radar replacing the nose machine-gun battery. In March 1945 the F7F-3 powered by the 2,300 hp R-2800-34W was introduced and this became the main production version with 189 being built. After the end of the War in August 1945 the F7F-3N was introduced as a night fighter version of the -3 and this could be recognised by a taller fin and more bulbous nose to house the radar. Sixty examples of the -3 were built, together with thirteen F7F-4Ns,

before production ended in late 1946. It is generally stated that
the -4N was the only Tigercat version equipped for shipboard
operations but this is not entirely true as all versions had
folding wings and most were fitted with arrester hooks. However, early carrier trials showed the need for several modifications, including a modified tail hook, strengthened airframe
and longer stroke undercarriage. It was only the F7F-4N that
actually incorporated all these improvements. Apart from the
aircraft itself, there were also ship compatibility problems as
the arrester gear of the early Essex class could not cope with
the landing weights of up to 20,000 lb and the twin-engine
tricycle undercarriage configuration was not compatible with
the standard deck barriers. Consequently, the Tigercat was
almost exclusively used by shore-based USMC squadrons.
Together with F4U-5N Corsairs, Marine F7F-3Ns provided an
important night interdiction capability in Korea, remaining in
service almost to the end of hostilities. However, apart from
occasional training exercises, the Tigercat never saw service
aboard the Navy's carriers. Interestingly two Tigercats were
supplied to Britain in 1946 as the Admiralty wished to gain
experience of the single-engine handling characteristics of
twin-engined carrier aircraft. The two F7F-1s were given British
serials (TT348 and TT349) while undergoing evaluation at
Farnborough before being returned to the US Navy in 1947.
Presumably these trials were in advance of the de Havilland
Sea Hornet entering service and also gave British pilots an
opportunity to assess the suitability of the tricycle undercarriage layout for carrier-based aircraft.

There was potentially one other twin piston-engined fighter
that might have seen service in the post-war era. This was
the highly unconventional single-seat Chance Vought XF5U-1,
which was built around a circular wing planform, giving it an
almost flying saucer-like appearance. Power was provided by
two 1,350 hp Pratt & Whitney R-200 Twin Wasp air-cooled
radial engines buried in the wing centre section, driving four-
bladed propellers mounted at the outboard edges by means of
extension shafts. The propeller blades were articulated to allow
a very high angle of attack to be achieved, giving an almost
helicopter like ability to fly at extremely slow airspeeds – a

feature that would have enabled the XF5U to have operated from the smallest carriers. Despite its unorthodox layout, the principle had been proven by the low-powered Vought V-173 research aircraft, which had flown in November 1942 and had demonstrated excellent handling characteristics. The XF5U prototype was rolled out in July 1946 but in the event was never flown and subsequently broken up. The projected performance included a maximum speed of 425 mph, a landing speed of only 40 mph, and a range of 710 miles. As well as being an unconventional design, the XF5U offered no advance over existing fighters and so it is not surprising that further development was not undertaken.

As already stated, the US Navy was well satisfied with the Hellcat, Bearcat and Corsair and consequently no other single-engined propeller-driven fighters were put into production in the latter stages of the War. However, mention should be made of one wartime prototype that was flown in 1944. This was the Curtiss XF14C-2, which was powered by a 2,300 hp Wright XR-3350-16 Cyclone turbosupercharged radial engine driving two three-blade contra-rotating propellers. Intended as a high-altitude interceptor, its performance was disappointing and certainly offered no improvement over the F4U Corsair. In addition, production of the Wright R-3350 engine was reserved at that time for the Boeing B-29 Superfortress and consequently the programme was cancelled.

The other aircraft to be found on the decks of American carriers at the end of the War were the Grumman TBF/TBM Avenger torpedo bomber and the Curtiss SB2C Helldiver. The Avenger, universally and affectionately known as the Turkey, had entered service in early 1942 and the elements of the first squadron (VT-8) were flown out to Midway Island at the start of June of that year, just in time to take part in the fierce Battle of Midway. Of the six aircraft that set off on the morning of 4 June to attack the Japanese carrier fleet, no fewer than five were shot down and the sixth was badly damaged, although the pilot managed to land back at Midway. Despite this inauspicious start, the Avenger went on to become a highly successful aircraft and was used in great numbers by the US Navy and also the Royal Navy. In all, some 9,839 Avengers were built, of which 2,293 were Grumman-built TBMs and the remaining 7,546 were General Motors (Eastern Aircraft) TBMs, the Royal Navy receiving 958.

The prototype XTBF-1 flew on 1 August 1941 and such was the rush to get this new aircraft into service, no fewer than 145 had been delivered by mid 1942, with production building up rapidly thereafter. Although of rather tubby appearance, the TBF Avenger was a typical sturdy Grumman-built aircraft well suited to the rough and tumble of carrier operations. Powered by a 1,700 hp Wright R-2600-8 radial engine, it could carry a torpedo or up to 2,000 lb of bombs in its enclosed bomb bay – this feature being a first for a carrier-based aircraft. A crew of three included a gunner who operated the 0.5 inch machine-gun in a dorsal turret set at the after end of the long cockpit canopy. There was also a ventral aft-firing 0.3 inch machine-gun with a limited field of fire and the pilot could operate a single nose-mounted 0.5 inch machine-gun. The top speed was a respectable 271 mph at 12,000 feet and the range was 1,215 miles. Avengers took part in all the major Pacific actions, being especially successful in the battles around Leyte Gulf when they were involved in the sinking of the large battleship *Musashi* and several carriers, and in April 1945 they achieved no fewer than ten torpedo hits on the other large battleship, *Yamato*, which was sunk as a result of this and other damage.

Part of the aircraft's success was that it proved to be tremendously adaptable in a variety of roles and the basic TBF/TBM-1 was produced in a number of specialised variants, most of which were radar equipped. Initially this was the ASB radar, which used wing-mounted Yagi antenna, but later versions carried the AN'/APS-3 or AN/APS-4 in a radome on the starboard wing. A few Avengers were modified as photographic reconnaissance aircraft with the installation of cameras in the bomb bay. By mid 1943 the addition of extra equipment had added almost 2,750 lb to the aircraft's empty weight so that performance and payload suffered accordingly. Consequently the XTBF-3 was flown in June 1943 powered by a 1,900 hp Wright R-2600-20 radial engine and this became standard in the TMB-3 production variant, which was delivered from April 1944. This variant was built exclusively by Eastern Aircraft and a total of 4,657 were produced before remaining production contracts were cancelled at the end of the War. Although no further aircraft were actually built, a prototype XTBM-4 was flown in April 1945 and had a strengthened airframe to enable it to withstand up to 5 g during dive-bombing attacks. It was planned to introduce this version to the production lines in August 1945 but this did not happen. The final variant would have been the TBM-5 in which a performance increase was sought by reducing the aircraft's weight. This was achieved by deleting the dorsal turret and reducing the crew to two, pilot and radio/radar operator. The latter could operate two rearward-firing 0.3 inch machine-guns on a flexible mounting (as opposed to a turret). Other changes included thrust augmentation exhaust pipes and a three-foot increase in wingspan to decrease wing loading. Two TBM-3 airframes were converted for trials beginning in June 1945 but, again, further work was stopped with the end of hostilities.

The basic TBM-3 was inevitably produced in several versions, of which the most important was the TBM-3E, the final production variant. This had a fuselage lengthened by almost a foot and carried an AN/APS-4 radar in a radome below the starboard wing. By dispensing with some non-essential equipment and the ventral 0.3-in machine-gun, the empty weight was reduced by some 300 lb despite the addition of the radar. The TBM-3E provided the basis for a number of sub variants, which were produced to meet the requirements of the post-war navy. Equipped with the basic torpedo bomber the Avenger squadrons were slowly run down, although many aircraft were transferred to Naval Air Reserve squadrons. In November 1946 the squadrons previously designated VT (Torpedo) were redesignated as VA (Attack) squadrons and at that time there were still eighteen Avenger squadrons allocated to the Carrier Air Groups. However, over the next three years these were re-equipped with the new Douglas AD-1 Skyraider (described below) and the Avenger finally gave up its frontline bomber role in 1949. However, some forty Avengers had been modified at the end of the war as night torpedo bombers with the dorsal turret removed to make room for a radar operator and his equipment and these aircraft served with Pacific and Atlantic

Fleet development squadrons (VCN-1 and VCN-2) into the early 1950s.

During World War II the Avenger was increasingly used in the ASW role and shared in the destruction of forty-two Axis, as well as one Vichy French, submarines. In the post-war era this role became even more important with the onset of the Cold War and Russian use of captured German technology to build a fleet of new submarines with a high underwater performance. Consequently a substantial number of Avengers were converted to TBM-3S configuration with all gun armament and the dorsal turret removed to make room for ASW and communications equipment. A searchlight was fitted under the port wing while a search radar was carried below the starboard wing. A sub variant was the TBM-3S2, which had a more advanced data link system installed.

The Avengers' stablemate aboard US carriers in 1945 was the Curtiss SB2C Helldiver. The US Navy had always been a staunch advocate of the dive bomber concept and this faith was reinforced when SBD Dauntless dive bombers from the carriers *Enterprise* and *Yorktown* scored a spectacular success, sinking three Japanese carriers at the very instant when it seemed that the battle was going disastrously wrong from the American point of view. Subsequently, further Dauntless attacks sank a fourth carrier, sealing the Japanese defeat. The Helldiver was intended as a replacement for the Speedy D (as the Dauntless was invariably called by Navy pilots) but it was never as well liked as its predecessor. The prototype XSB2C-1 flew on 18 December 1940 powered by a 1,700 hp Wright R-2600 Double Cyclone radial engine, giving it a range of 1,330 miles at a cruising speed of 244 mph. The armament comprised two fixed forward-firing 0.3 inch machine-guns and two 0.5 inch machine-guns on a flexible mounting in the rear cockpit. A single 1,000 lb bomb was carried in an internal bomb bay with a crutch mechanism, which ensured that the bomb fell clear of the propeller arc when released in a vertical dive.

The prototype had an unhappy history, suffering major damage in an accident in February 1941. Although repaired, it was subsequently destroyed after an inflight wing failure in December 1941. However, the test programme revealed serious handling problems and consequently the fin and rudder were considerably enlarged. Other changes for production aircraft included self-sealing fuel tanks and the fixed armament was increased to four wing-mounted 0.5 inch machine-guns. However, the empty weight rose by nearly 40 per cent, from 7,030 lb to 10,114 lb, and this had a detrimental effect on performance, the maximum speed falling from 325 mph to 273 mph, while the service ceiling was only 21,000 feet compared with the prototype's 30,000 feet. The first production SB2C-1 flew on 30 June 1942 but was subsequently lost, again due to wing failure in a high-speed dive, in January 1943. By that time the first operational squadrons were forming but with significant flight restrictions and the first operational sorties were not flown (by VB-17 against targets at Rabaul) until November 1943. Over the following two years the Helldiver replaced the Dauntless aboard the Pacific Fleet's carriers and the final production versions were the SB2C-4 and -5. These had a more powerful R-2600-20 engine, giving 1,900 hp, and the offensive load was increased to include eight 5 inch rockets or two 500 lb bombs carried underwing in addition to the internal 1,000 lb bomb. Perforated dive brakes on the wing trailing edge improved handling in the dive. In all, Curtiss delivered 4,105 Helldivers and another 1,294 were built under licence in Canada by Fairchild, and Canadian Car and Foundry under the designations SBF and SBW respectively. Twenty-six SBWs were supplied to the Royal Navy who did not use them operationally. Any outstanding orders were cancelled at the end of the War and no further development was undertaken, although at that time two XSB2C-6 Helldivers were undergoing trials. These had lengthened fuselages and squared wingtips in an effort to improve directional stability, which remained a problem throughout the aircraft's career. Post-war service was limited as the dive bomber squadrons were run down in favour of the new attack squadrons and by the time of the Korean War no Helldivers remained in frontline service.

The aircraft that replaced the Avengers and Helldivers in the post-war period were the new breed of attack aircraft. With the destruction of the Japanese Navy, the US Navy would no longer expect to fight major sea battles. Even with the onset of the Cold War, the Soviet power was vested in its land and air forces and its naval strength was initially very limited. Consequently, US Navy doctrine called for the ability to land and support forces ashore, and to launch heavy attacks on shore targets. Thus even by the end of the War the requirement was for aircraft that could offer a good performance while carrying a heavy offensive load. In fact, aircraft such as the Avenger and Helldiver actually carried no more than fighter-bombers such as the Hellcat and the Corsair while having a poorer performance. This was partly due to the fact that such aircraft carried additional crew members and defensive armament, which increased weight and lowered performance. The increased availability of electronic navigation aids such as radar and TACAN meant that the requirement for a navigator was much reduced and single-seat high-performance aircraft were capable of acting as bombers. Those aircraft that began to enter service from 1945 onwards resulted from specifications drawn up with these ideas in mind, although initially the multi-crew aircraft held sway.

As far back as late 1941, the US Navy was already looking for aircraft that would eventually replace the Helldiver and this was to produce a rash of prototypes, although eventually only one was to achieve production status. One of the earliest of these projects was the Douglas XSB2D-1, a powerful two-seat bomber intended as an eventual replacement for the Dauntless. In this context it is interesting to note that the US Navy was continually looking ahead, and almost inevitably as one aircraft type entered service plans were already being drawn up for a successor. The Douglas design was a great advance on its predecessor, both in terms of performance and capability and introduced a number of novel features. These included a tricycle undercarriage, a laminar flow wing and a defensive armament consisting of remote-controlled turrets. A gull wing configuration, with the outer sections folding upwards for stowage, left room for a capacious internal bomb bay (a similar arrangement was adopted for the British Blackburn B-54/YA.7, which was developed in the late 1940s). Powered by a 2,300 hp Wright R-3350-14 air-cooled radial engine, the prototype XSB2D-1 flew for the first time on 8 April 1943 and demonstrated an excellent performance. The top speed was almost 350 mph and it could carry an offensive load of up to 4,200 lb – twice that of the contemporary Curtiss Helldiver. This could consist of two 1,600 lb bombs in the bomb bay and two 500 lb bombs on underwing racks, or two 2,100 lb torpedoes carried externally beneath the fuselage. The crew consisted of a pilot and radio operator/gunner, the latter controlling dorsal and ventral turrets, each carrying two 0.5 inch machine-guns, while two wing-mounted 20 mm cannon were proposed. By August 1943 Douglas had started production to meet orders for a total of 358 SB2D-1s, as well as the two prototypes. Unfortunately, by this time the US Navy had reviewed its policy in respect of strike and attack aircraft, experience in the Pacific indicating that a defensive armament was unnecessary given the increasing degree of air superiority enjoyed by the Fleet's carrier air groups. Consequently, in late 1943 and early 1944 the Bureau of Aeronautics issued specifications and contracts for a series of prototype single-seater carrier borne bombers.

In addition, Douglas was asked to look at modifying the SB2C-1 design as a single-seater and this resulted in the BTD-1 Destroyer, which flew in prototype form on 5 March 1944. This aircraft was actually the second production SB2D-1 and the required modifications were relatively straightforward. The remotely controlled turrets were removed, a shortened cockpit canopy was provided for the pilot and the tail fin was extended by means of a large dorsal fillet. Weight saved by these changes was usefully employed to increase armour protection and fuel tankage. The performance and offensive load were unaltered and the wing-mounted 20 mm cannon were retained. The prototype was followed by twenty-seven production BTD-1s, the last being delivered on 8 October 1945. Although the switch to production of the single-seat version had been achieved quite easily, the BTD Destroyer inevitably suffered in comparison with other designs in view of the fact that it was an adaptation of an earlier two-seater aircraft. Consequently, production ceased after the initial batch and the type never entered operational service, although various examples were utilised for tests and trials in the immediate post-war era. Of these the most interesting were two XBTD-2s, which, as will be related, played an important role in the introduction of jet aircraft by the US Navy. Also, and perhaps more importantly, this did not mean the end of the Douglas involvement in the competition to provide a new single-seat bomber for the US Navy.

In the meantime Douglas had built and flown the proto-type of yet another large multi-crew torpedo bomber. This resulted from a US Navy contract issued in December 1943 for a torpedo bomber to operate from the decks of the new Midway class carriers then being laid down. The resulting XTB2D-1 Devastator II (later renamed Skypirate) was by far the largest and heaviest carrier-based aircraft at the time of its maiden flight on 13 March 1945, with an all-up weight of 28,000 lb and a high aspect ratio wing with a 70-foot span. Power was provided by a 3,000 hp Pratt & Whitney XR-4360 driving a complex eight-bladed contra-rotating propeller assembly. The planned armament include four wing-mounted 0.5 inch machine-guns, twin 0.5 inch guns in a dorsal turret and a single ventral 0.5 inch gun. Up to 8,400 lb of bombs and torpedoes could be carried. Despite this formidable load, the Skypirate was a sprightly performer with a maximum speed of 340 mph at 15,600 feet and a normal range of 1,250 miles, extended to 2,880 miles with extra tankage. In terms of effectiveness it could carry two torpedoes almost 150 miles further than an Avenger could carry one. Despite this, the Skypirate also fell victim to the US Navy's policy of adopting single-seat bombers and development was cancelled at the end of the War.

While the single-seat Douglas BTD-1 was being flown, the US Navy had contracted for prototypes of a number of other single-seat bombers, these resulting in the Curtiss XBTC-1, Kaiser Fleetwings XBTK-1 and Martin XBTM-1. Curtiss, of course, was keen to emulate its success with the SB2C Helldiver and in December 1943 received a contract for two prototype XBTC-1 torpedo bombers. However, the first of these did not fly until July 1946, by which time the decision to fit an improved version of the Pratt & Whitney XR-4360-8A Wasp Major resulted in the designation XBTC-2. With an output of 3,000 hp driving a six-bladed contra-rotating propeller, this big engine provided excellent performance with the top`speed being 374 mph at 16,000 feet and the maximum range being 1,835 miles, although the service ceiling was only just over 26,000 feet. The armament comprised four wing-mounted 20 mm cannon but the offensive load was only 2,000 lb of bombs or a single torpedo. In practice the performance figures were irrelevant as by mid 1946 other rival designs were already in production, adequately meeting the US Navy's requirements, and no further development of the XBTC-1 was undertaken.

Another single-seat bomber that was produced only in proto-type form was the little known Kaiser-Fleetwings XBTK-1, which flew in April 1945 powered by a 2,100 hp Pratt & Whitney R-2800-34W air-cooled radial engine. The top speed was around 342 mph at sea level and a single torpedo or up to 5,000 lb of bombs could be carried while the fixed armament comprised two wing-mounted 20 mm cannon. Only two prototypes were flown and a third airframe was used for structural testing. No production orders were forthcoming. However, more success met the third contender, the Martin XBTM-1, which first flew on 24 August 1944. Powered by a 3,000 hp Pratt & Whitney XR-4360-4, the XBTM-1 had a respectable performance with a top speed of 367 mph at 11,000 feet and a range of 1,800 miles. More importantly, it could lift up to 4,500 lb of ordnance, carried on no fewer than fourteen underwing and one under fuselage hardpoints, in addition to a fixed armament of four 20 mm cannon. In fact, on one demonstration flight, a Mauler flew with a 10,689 lb ordnance load including no fewer than three torpedoes! Flight testing revealed no major problems and in January 1945 an order for 750 aircraft was placed. Before the first of the production aircraft flew in December 1946 the US Navy had redesignated the BT torpedo bombers as A attack aircraft. As a result the XBTM-1 became the AM-1 and the name Mauler was adopted. After carrier qualification trials in 1947, the first operational unit, VA-17A, was formed in March 1948 and other units followed. However, production of the Mauler ceased in October 1949 after a total of 151 had been produced (including prototypes). Subsequently it was withdrawn from frontline use and passed on to Reserve squadrons, which continued to fly Maulers for a few more years.

The reason for the premature ending of the Mauler's career was due to the unprecedented success of the rival Douglas

AD-1 Skyraider. This versatile and long-serving aircraft had its origins in the failure of the company to gain production contracts for the BTD-1 Destroyer or the XTB2D-1 Skypirate. In July 1945 the Douglas design team, led by Chief Engineer Ed Heinemann, famously got together in a hotel room and between them thrashed out the basic characteristics of a new single-seat attack bomber. Experience with work already carried out with the earlier designs was incorporated and the resultant proposal impressed the Bureau of Aeronautics enough to authorise the construction of no fewer than fifteen prototype and pre-production aircraft to be designated XBT2D-1 Destroyer II. Douglas was under considerable pressure to complete the design work and get a prototype flying as it was starting six months later than its competitors. In fact, the first flight of the prototype XBT2D-1 took place on 18 March 1945, four months ahead of the planned schedule and well under twelve months from that momentous hotel room meeting – a remarkable achievement by any standards. The company's efforts were rewarded when the aircraft sailed through its test and evaluation programme at the Patuxent River Naval Test Station in only five weeks with glowing reports from the pilots involved. Douglas immediately received a contract for 548 BT2D-1s in May 1945 but this was reduced to an initial order for 277 aircraft after the end of the War.

The Douglas BT2D-1 was of conventional configuration with a low-mounted, straight, tapered wing and a slab-sided rear fuselage with dive brakes fitted on either side. The pilot sat well forward under a raised clear-view bubble canopy, giving excellent forward visibility despite the large Wright R-3350-24W radial engine, which developed 2,500 hp (the first four aircraft were fitted with 2,300 hp R-3350-8 engines). This enabled the prototypes to achieve maximum speeds in excess of 350 mph while up to 8,000 lb of ordnance could be carried on three large pylons, one under each wing and another on the centreline, supplemented by six smaller hardpoints on each wing for small bombs or rockets. Production aircraft began to reach the US Navy in early 1946 when the aircraft's designation was changed to AD-1 to conform with the new attack category and the name Skyraider was officially adopted. At this stage some problems associated with the aircraft's rushed development became apparent and these included propeller vibration, power-plant installation defects and several undercarriage failures. Rectification of these problems meant that the first Skyraider squadron, VA-19A, did not become operational until the end of 1946 but thereafter the other squadrons quickly re-equipped with the new bomber, which replaced Maulers, Avengers and Helldivers.

The inherent flexibility of the basic design led to numerous developments and there were eventually eight major variants produced in no fewer than thirty-seven sub versions. The Skyraider remained in production until February 1957, by which time some 3,180 had been delivered, and it remained in front-line US Navy service until 1968, flying offensive sorties against Vietnam targets from the USS *Coral Sea*. In addition to US service, the Skyraider was also supplied to Britain, France and South Vietnam, the latter using the aircraft in action until the end of the Vietnam War in 1975. In fact, serious consideration was given to re-opening the production line in 1965 as the Skyraider was proving so useful in the gathering Vietnam conflict, although practical difficulties resulted in this plan being dropped.

To go back to the Skyraider's early days, a number of the XBT2D-1 prototypes were converted to specialised roles. These included the -1N night attack bomber with two extra crew members accommodated in the fuselage and radar and search-light pods carried underwing. The XBT2D-1Q was an electronic countermeasures aircraft with the operator again housed in the fuselage and radar and ECM equipment in underwing pods. This last variant was so successful that thirty-five were ordered as the AD-1Q in addition to orders for 242 AD-1s, the standard single-seater attack version. Another early version was the XAD-1W, which had a large radar antenna in an under fuselage radome and two radar operators accommodated under a fairing behind the pilot's cockpit. Following successful trials, a total of 417 AEW Skyraiders were built as the AD-3W, AD-4W and AD-5W. The next major variant was the AD-2, which had a more powerful 2,700 hp R-3350-26W, a strengthened airframe and additional fuel tankage. The maximum all-up weight rose to 18,263 lb compared with the prototype's 17,500 lb. Only 156

AD-2s were built together with 125 AD-3s, which featured incremental improvements and an MAUW of 21,000 lb.

The AD-4, which began to leave production lines in 1949, was built in greater numbers than any other Skyraider variant and no fewer than 1,051 were built, being produced or modified to result in eight versions. Apart from the basic AD-4, these included the AD-4W AEW version and the AD-4B nuclear bomber. Although originally produced with only two wing-mounted 20 mm cannon, most AD-4s (except for the unarmed AD-4W) were later fitted with four cannon, which became standard on subsequent production versions. It was a standard AD-3B that set a load-carrying record for single-engined aircraft on 21 May 1953 when it flew with a load consisting of three 1,000 lb, six 750 lb and six 500 lb bombs. Including the weight of the pylons, racks and fuel, the total came to 14,941 lb and although not representative of loads carried under combat conditions, it did illustrate the amazing capabilities of the Douglas bomber.

In late 1948 Douglas proposed a new and enhanced Skyraider provisionally designated AD-5 powered by a turbo-compound version of the R-3350 but this was not flown. Neither was an ASW hunter killer version proposed in 1949 and also again provisionally designated AD-5. By mid 1950 it appeared likely that production of the basic Skyraider (866 had been completed by May 1950) was coming to an end as the US Navy's peacetime requirements were met. However, the Communist invasion of South Korea on 25 June 1950 completely altered the picture and a requirement for close support and ground attack aircraft was quickly established. While the new jet fighters could provide air superiority, the ground attack mission centred on load-carrying ability, loiter time over target, and resistance to ground fire. On all of these counts the piston-engined bomber still held many advantages and consequently development of an updated version of the Skyraider was pressed ahead. The AD-5 proto-type, which first flew on 17 August 1951, embodied many lessons learnt from earlier production and was intended as a multi-role two-seat aircraft. Apart from the provision of a widened forward fuselage to accommodate the two crew, it was also lengthened by two feet and the vertical fin was increased in size as compensation. The standard armament was four wing-mounted 20 mm cannon and the AD-5 was produced in three versions – a basic AD-5 (212 built), a night attack AD-5N (239 built) and AEW AD-5W (218 built). There was also a single experimental ASW AD-5S equipped with a magnetic anomaly detector and fifty-four AD-5Ns were modified to AD-5Q standard as four-seat electronic countermeasures aircraft. Subsequently Douglas built 713 AD-6 Skyraiders, which reverted to single-seat configuration but otherwise incorporated all of the changes introduced by the AD-5. The final production version was the AD-7, which was similar to the -6 but had strengthened wings and undercarriage. The last of these was delivered in February 1957.

In Korea the Skyraider quickly established an outstanding reputation as a ground attack aircraft, its ability to carry a heavy and varied ordnance load being particularly welcomed by troops on the ground. To meet demand the aircraft was produced in ever increasing numbers, although the peak rate of fifty-nine aircraft in one month was not reached until June 1954, almost a year after the Armistice was signed in July 1953. Despite the ever increasing number of jets coming into service (as well as the flight of the Douglas A4D lightweight jet bomber in 1954) the peak US Navy and Marine Skyraider strength was not reached until 1955 when no fewer than twenty-nine US Navy and thirteen Marine squadrons operated the versatile bomber. The subsequent career of the Skyraider lies outside the scope of this book but it should be recorded that the AD (or Spad as it was affectionately known) remained in frontline service for more than another decade, principally as a result of the US involvement in the Vietnam War. However, it should also be noted that the Royal Navy took delivery of fifty AD-4W AEW aircraft under MDAP arrangements. These were operated from 1952 until 1960 by 849 Squadron, which acted as parent unit for flights of four aircraft detached to operational carriers.

The success of the Skyraider removed any incentive for the US Navy to consider further development of piston-engined attack aircraft and future requirements were to met by jet aircraft. However, the Skyraider did act as a basis for the turboprop-powered Douglas A2D Skyshark, which is described in Chapter 8.

3

CARRIER FLEETS – 1945 TO 1950

At the end of the War in 1945 the Royal Navy was no longer the world's most powerful navy, having ceded that position to the US Navy. Nowhere was this more clearly illustrated than in a comparison of carrier strength. Whereas in 1939 the US Navy had only five carriers in commission, the Royal Navy had seven, although a review of aircraft numbers would actually have weighed well in favour of the American ships. However, by August 1945 the situation had changed out of all recognition and the US Navy was able to call on over twenty large fleet carriers and nine smaller light fleet carriers, as well as numerous escort carriers. On the other hand the core strength of the Royal Navy lay in six large fleet carriers with armoured decks and four new light fleet carriers, which had just come into service but did not see any wartime action. There were also substantial numbers of escort carriers, although the majority of these were US built. One point that the carriers of both navies had in common was that in most cases their basic designs went back to the pre- or early war years. Although most had been modified in the light of war experience, they were not well equipped to deal with the new generations of heavier and faster aircraft that would enter service in the post-war decade.

The build up of the US Navy's carrier force had been driven by the need to build and equip a fleet capable of taking the war across the vast Pacific Ocean to the shores of the Japanese homeland and in this task they were ably supported by the massive US industrial infrastructure. Once geared up for war production, the US shipyards were able to turn out warships of all types in numbers that the British could only envy and the

Axis partners could never hope to match. By 1945 many of the American carriers that were in service immediately after Pearl Harbor had been lost in action in the fierce carrier battles of 1942 and the only survivors in frontline service were the 19,800-ton USS *Enterprise* and the 33,000-ton USS *Saratoga* (the smaller USS *Ranger* also survived but had seen little action and was used for training). In a desperate effort to reinforce the fleet, nine Cleveland class cruiser hulls were adapted to form the basis of an 11,000-ton light fleet carrier (Independence class) and these all entered service in 1943. These provided a welcome boost to offset the earlier losses but the most significant new carriers were those of the 27,500-ton Essex class, which were eventually to provide the backbone of the US carrier fleet not only in the latter half of World War II but also for the subsequent decades and the period covered in this book.

The Essex class had its origin in the mid 1930s when studies were made on the characteristics to be incorporated in any future carrier. Under the terms of the Washington and London Naval treaties, the size of carriers was restricted to a maximum of 27,000 tons and there were limits on the total tonnage of carriers permitted to the signatory navies. However, by 1937 these treaties had lapsed or were in abeyance and consequently the US Navy was free of any artificial constraints when considering various design studies for the new carriers. Priority was given to embarking the maximum number of aircraft together with all the facilities required for their efficient operation. The operational requirement was for four squadrons, each of eighteen aircraft, with space for reserve aircraft and spares. The design was based on experience with the previous smaller Yorktown class but the overall dimensions were similar to the much larger *Lexington* and *Saratoga*, which had been produced by converting battlecruiser hulls. The flight deck was 886 feet long and almost 90 feet wide and in the American tradition was wood planked, although underneath was light armour plating. Below this was the hangar deck measuring 580 feet by 71 feet and having no less than 18 feet of headroom. The hangar was served by two deck centreline lifts situated fore and aft, and a deck edge lift on the port side amidships (this could be folded up to allow transit of the Panama Canal). These arrangements enabled these ships to operate with a normal complement of eighty aircraft, although in the closing stages of the war this was increased to around 100 at times.

As well as the aircraft, the Essex class carriers were well equipped for defence against air attack with a battery of twelve 5 inch/38 cal dual-purpose guns in four twin and four single mounts, while a battery of no fewer than around forty-four 40 mm and another forty-four 20 mm AA guns provided short-range defence. This was actually less than the designed light AA armament and the numbers of guns carried was increased substantially at a later stage to cope with the threat posed by Japanese kamikaze attacks. Unlike British carriers, the Essex class was very lightly armoured and relied on an extensive system of watertight compartments to absorb the effects of bomb and torpedo attacks. In practice this worked well in terms of ship survivability (no Essex class carrier was sunk) but almost inevitably any bomb hits caused substantial damage to the flight deck and could put the ship out of action for months. The most dramatic example was the USS *Franklin* whose crowded flight deck was hit by two bombs in March 1945. Rapidly spreading fires and blast damage killed no fewer than 724 men and injured another 265. Despite this the fires were eventually brought under control and the ship was able to return stateside for repairs under her own steam, although she saw no further action.

Perhaps the most amazing aspect of the Essex class is the speed at which the carriers were produced and the numbers actually built. The name ship was laid down in April 1941 and commissioned on 31 December 1942. By the end of 1943 another six were in commission and these ships were forming the spearhead of the Task Forces then beginning to work their way across the Central Pacific. Another seven commissioned in 1944 and by August 1945 there were seventeen in service with two more commissioning, too late to see service, before the end of the year. Despite the coming of peace, a further four were completed in 1946 bringing the total to twenty-three and one more, the USS *Oriskany*, was completed to a modified design in 1950. Thus in the post-war era the US Navy had the advantage not only of having a large force of fleet carriers but also gained operational efficiency from the fact that they were all built to the same basic design. This allowed for considerable standardisation in equipment and operational procedures.

Of the other carriers in service in 1945, the pre-war survivors (*Ranger, Enterprise* and *Saratoga*) were quickly de-commissioned, as were most of the Independence class. Of the latter, two were lent to France (*Belleau Wood* and *Langley*) and another (*Cabot*) eventually sold to Spain. Although a few were used for training purposes at various times, only one (USS *Bataan*) saw frontline service as an attack carrier off Korea between 1950 and 1953. During this time she embarked USMC squadrons operating F4U Corsairs.

Despite the success of the Essex class carriers, the US Navy had three much larger carriers under construction in 1945. These were the 45,000-ton Midway class of which the name ship commissioned on 10 September 1945, only a few days after hostilities ended. The second, USS *Franklin D Roosevelt*, commissioned only a few weeks later. Completion of the third ship, USS *Coral Sea*, was not hurried and she was not completed until October 1947. In essence these ships were scaled-up Essex class with the aircraft complement increased to over 130, these being flown from a flight deck measuring 932 feet by 137 feet. Like the Essex class they carried a heavy AA armament comprising eighteen 5 inch/38 cal DP guns, all in single mountings along either beam below flight deck level. Provision was also made for eighty-four 40 mm and eighty-two 20 mm light AA guns, although this full complement was never fitted to any of the trio.

The increased size of the Midway class was not actually a result of any desire to allow the operation of more aircraft but was dictated by evaluation of the early war experience of the British armoured carriers in the Mediterranean where both *Illustrious* and *Formidable* survived several hits by 1,000 and 2,000 lb bombs (amazingly, *Illustrious* took a total of ten hits). This was in stark contrast to subsequent American experience in 1942 and this prompted the US Navy General Board to insist on an armoured deck for the new carriers then being planned. In the American system, the flight deck was not an integral part of the structure of the ship but was basically a platform supported above the main hull structure. This facilitated features such as open hangar sides and deck-edge lifts, but did not contribute to the strength of the ship and posed problems when the weight of a 3.5 inch armoured deck was added. By contrast, in the British armoured carriers the flight deck and enclosed hangar walls formed a rigid box girder, which contributed substantially to the strength of the whole ship. As well as protection against air attack, the Midway class was also designed to be armoured against 8 inch gunfire, which resulted in an 8 inch waterline belt, armoured bulkheads and more armour for the main and hangar decks. The total weight of armour amounted to some 4,000 additional tons and this naturally had an adverse effect on speed. In order to maintain a fleet speed of 32 knots more powerful machinery was required, which took up additional space, leading to an increase in overall dimensions. Thus the increased flight deck area and hangar space was a fortunate and useful by-product of the need to produce an armoured version of the Essex class. Deck arrangements were similar to the previous ships with two centreline lifts fore and aft, and a large deck edge lift on the port side. Unlike the Essex class this could not be folded to the vertical position as the new ships were too wide to transit the Panama Canal.

Original plans called for six Midway class carriers but one order was cancelled in November 1943 and the other two in March 1945 when it was clear that they would not be needed for wartime service. In the immediate post-war era the Midway class with their large flight decks were ideal for the trials and early deployment of the first jets. Indeed, the first landing by a US Navy jet aboard a carrier took place aboard the USS *Franklin D Roosevelt* on 21 July 1946 and subsequently the first US Navy jet fighter squadron (VF-17A) made an initial deployment aboard the ship in 1947.

In Britain, the Royal Navy was not able to enjoy the luxury of a large fleet of newly built carriers and for a while work on carriers under construction virtually came to a halt as soon as hostilities ended in August 1945. The six operational fleet carriers had all seen arduous war service and most were in need of a refit. A further problem was that they lacked the homogeneity of the American Essex class, being in fact made up of three distinct sub groups. The first three ships (*Illustrious, Formidable* and *Victorious*) had all been completed to the original design with an armoured flight deck and a single deck hangar capable of accommodating thirty-six aircraft. These entered service in 1940/41 and at the end of 1941 were joined by a fourth ship, HMS *Indomitable*. In this ship the design was recast to increase hangar capacity to forty-eight aircraft and this was done by inserting a lower hangar deck below the after section of the main hangar deck. However, whereas the main hangar in the original ships had a headroom of 16 feet, this was reduced to 14 feet in *Indomitable*, although the lower half hanger retained the 16-foot clearance.

The next two ships, *Implacable* and *Indefatigable*, were laid down in 1939 but not completed until 1944 (compare this with the average of around eighteen months or less to complete some of the American Essex class carriers – a clear demonstration of the strength of the US industrial base). In this case the lower hanger deck was extended to a total length of 208 feet but headroom in both hangars was only 14 feet and in practice the forward section of the lower hangar was taken over to provide additional storage and accommodation. As a result the designed aircraft complement was only slightly increased to fifty-four.

All six of these carriers were heavily armed with a total of sixteen 4.5 inch DP guns in semi-recessed twin mountings positioned in pairs at the forward and after corners of the flight deck. Secondary AA armament varied but by the end of the War most ships carried the original forty-eight 2-pdr guns in quadruple mountings backed up by around forty to sixty 20 mm guns in single or twin mountings (in many cases any twin 20 mm mountings had been replaced by single 40 mm guns in ships deployed in the Pacific). This heavy armament, together with the armoured deck and hangar, was one of the reasons why the British carriers carried significantly fewer aircraft than their American contemporaries. In fact, Royal Navy doctrine at the start of the War envisaged all aircraft being struck down into the hangar in the event the ship came under attack, defence then being left to the guns and escorting vessels. It was only slowly realised that high-performance fighters were the real answer to air attack and demand for these grew throughout the War. Eventually the Royal Navy adopted the American concept of a permanent deck park and was able to increase the numbers of aircraft carried. The first four ships could eventually embark up to fifty-four aircraft while in *Indomitable* and *Indefatigable* this rose to eighty-one by a combination of a deck park and the use of outriggers. (Outriggers were narrow tracks projecting from the edge of the flight deck in which the tailwheel could rest while the aircraft was pushed back so that it overhung the deck edge – precarious but effective!)

Despite the rise in aircraft complement the latter two ships were handicapped by their hangar headroom and could only accommodate aircraft whose wings folded alongside the fuselage (such as the Hellcat and Avenger) but not the larger Corsair, which otherwise became the standard Royal Navy fighter in the Pacific. The British Seafire could also be accommodated in the lower hangers as although its wings folded upwards, the tips then folded down to reduce height. In this respect the light fleet carriers of the Colossus class that were coming into service in 1945 were a much better bet with no less than 17.5 feet of headroom in their single hangar. While the varying hangar heights may have been inconvenient at worst in the closing stages of the War, it became much more significant in the post-war era as the size of aircraft increased. In fact, despite being the newest and least damaged of the fleet carriers, the two Implacable class had a very limited post-war career, being laid up in reserve or used as training ships. When the time came to consider the modification and modernisation of existing ships to cope with new generations of post-war aircraft, the work involved in either increasing hangar headroom or converting to a single hangar deck was just too expensive to contemplate and these fine ships were scrapped in the mid 1950s.

By 1942, at the same time that the US Navy was ordering the Midway class, the Royal Navy was also laying down the first of four planned new carriers of the Audacious class, which would benefit from wartime experience. The basic design of the Implacable class was followed but the lower hangar was extended so that it was served by both the forward and after centreline lifts (in all previous ships the lower hangar was only served by the after lift – a potential problem if this was put out of action for any reason). A 4 inch armoured deck was planned and a complement of seventy-eight aircraft would be embarked. Initially the hangar height was set at 14.5 feet but by the time orders were being placed, it was realised that this was too low to accommodate the American aircraft on which the Royal Navy was becoming dependent and so this was increased to 17.5 feet. This had several knock-on effects and the increase in hull depth by 6 feet required a substantial increase in beam to maintain stability. The increase in dimensions extended to the hangars where more aircraft could be stowed and the complement rose to 100, requiring additional stores and fuel to be stowed. Consequently the standard displacement rose from 24,000 tons for the initial design to 33,000 tons in the revised version and, in fact, when the first ship (HMS *Eagle*) commissioned in 1951 she displaced 36,800 tons (45,750 tons at full load).

Of the original four ships, one was cancelled before construction began and a second was cancelled in October 1945. The lead ship, *Audacious*, was renamed *Eagle* and as noted was completed in 1951. Her sister ship, *Ark Royal*, was not completed until 1955, by which time she had been extensively modified and incorporated several new features to enable her to operate new-generation jet aircraft, as will be described later. Had the War not ended, there were plans for three further carriers of the Malta class, which were ordered in 1943, although all were cancelled at the end of 1945 when only preliminary work on the lead ship had commenced. Whereas the *Eagle* and *Ark Royal* were really only just equivalent to the American Essex class in terms of flight deck size and aircraft capacity, the Malta class would have been more in line with the Midway class with similar flight deck dimensions (900 × 136 feet). The standard displacement at 46,900 tons was also similar but whereas the new American carriers adopted the British principle of an armoured flight deck, the Royal Navy decided to abandon this feature and instead followed previous American practice of a lightly armoured flight deck and open hangar structure. Two deck edge lifts were incorporated (a first for a British carrier at that stage) as well as two centreline lifts. The standard battery of sixteen 4.5 inch DP guns was retained and this was to be backed up by over fifty 40 mm AA guns. The stated aircraft complement was set at only eighty aircraft but this was set by hangar capacity and would have been increased by the use of deck parks.

Despite the orders placed for the Audacious and Malta classes in 1942 and 1943, only *Eagle* and *Ark Royal* were actually completed and as already related, these did not enter service until 1951 and 1955 respectively. This meant that British carrier capacity up to the outbreak of the Korean War in 1951 rested on the Illustrious/Indefatigable class fleet carriers and the ten smaller light fleet carriers of the Colossus class, which were completed between 1944 and 1946. To take the fleet carriers first, some of these saw little or no use as operational carriers before being scrapped in the 1950s. HMS *Formidable* was laid up in reserve in 1947 after several voyages to the Far East acting as a troopship. Although subsequently considered for modernisation, inspection revealed that she was in too poor a condition and she was scrapped in 1953. HMS *Victorious* was also used for training duties and as a training ship until laid up in 1950 for an extensive eight-year modernisation refit (described later).

Paradoxically, it was the oldest of the fleet carriers, HMS *Illustrious*, which saw the most active service in the post-war era. This was partly because in 1945 when the hostilities ended she had just commenced a major refit and the opportunity was taken to incorporate modifications that would enable her to operate some of the newer aircraft then about to enter service. The forward and after ends of the flight deck were extended and widened, additional arrester wires fitted and the deck lifts were enlarged and strengthened. Considerable efforts were made to improve crew accommodation and facilities, restoring recreational spaces that had previously been taken over for other purposes under wartime manning conditions. The light AA armament was rationalised and comprised nineteen single 40 mm and twenty single 20 mm guns. The original multiple 2-pdr mountings were retained. The radar outfit was updated and of particular interest was the installation of a Carrier Controlled Approach (CCA) radar and closed circuit television cameras to allow flight deck operations to be monitored. Both of these were firsts for a British carrier. These modifications made *Illustrious* the Royal Navy's most up to date carrier and it was natural therefore that the ship would be utilised on several occasions for tests and trials with new aircraft. In 1947 for instance, the prototype Supermarine Attacker, which was to be the Royal Navy's first fully operational jet fighter, carried out deck landing trials. A further refit in 1948 resulted in the multiple 2-pdrs being replaced by twin 40 mm guns and the radar outfit was again improved. From the aviation point of view the catapult was modified to permit the launching of heavier aircraft. Subsequently *Illustrious* remained in service as the Home Fleet trials and training ship until she was laid up in 1954. During that time virtually every prototype naval aircraft including the jet-powered Sea Hawk and Sea Venom, and turbo-prop Gannet and Wyvern flew from her decks. In addition a whole generation of naval pilots made their first deck landing aboard *Illustrious*.

The fourth ship of the class, HMS *Indomitable*, also saw some significant post-war service, although initially she was employed on trooping duties before being taken in hand for a three-year modernisation refit in 1947. In general this followed the changes already incorporated in HMS *Illustrious* and in 1951 she joined the Home Fleet as flagship and as a fully operational carrier with an air group comprising Sea Furies and Fireflies, with Firebrands and Sea Hornets being embarked on occasions. One significant feature was that *Indomitable* was the first British carrier to permanently embark a helicopter, this being a Westland-built S-51 Dragonfly, which was used for search and rescue duties. *Indomitable*'s operational career ended in October 1953 and she was scrapped two years later.

Despite being the newest fleet carriers, *Implacable* and *Indefatigable* were, as has already been noted, handicapped by the low headroom of their hangar decks. In fact, *Indefatigable* was used almost exclusively as a troop ship and then as a sea-man's training ship, the latter role involving the construction of classrooms on her flight deck, which precluded any aviation activity, and she was decommissioned in 1955. On the other hand, *Implacable* saw a more active post-war career. She did not actually return to the UK after the end of the War until

June 1946, having provided an operational presence in the Far East with an air group of Seafires later joined by Avengers (828 Squadron, the last to operate the torpedo/bomber version of this aircraft) and Fireflies. She then recommissioned with the Home Fleet and in March 1949 she embarked 801 Squadron equipped with Sea Hornet F.Mk.20s and Firebrand TF.Mk.5s – the first operational deployment of these aircraft. Later that year she embarked four de Havilland Sea Vampire F.Mk.20s and these took part in exercises off Gibraltar, this being the first operational deployment of jet fighters by the Royal Navy, although four years had elapsed since a Vampire had made the first landing of any jet aboard an aircraft carrier in December 1945. Although in the forefront of naval aircraft development, *Implacable* was also involved in something of retrograde step when in 1950 she embarked 815 Squadron equipped with Fairey Barracudas modified for anti-submarine use. This came about due to a chronic shortage of ASW aircraft pending the introduction of the turboprop Gannet, although this would not occur until 1955. *Implacable* was paid off into reserve in September 1950 and subsequently recommissioned in 1951 as a seaman's training ship and although in this the capability to operate aircraft was retained she was never again used as an operational carrier. She was paid off in August 1954 and was scrapped in the following year.

In this way the wartime fleet carriers rapidly faded from the scene and, with the exception of *Victorious* (whose further career is described in Chapter 5), they had all been laid up or scrapped within the decade covered by this book. As will be seen, this is in stark contrast with the American Essex class, which were constantly updated and remained in service well into the 1970s. There were a number of reasons for this, including the state of the ships after arduous wartime service and the limitations imposed in some cases by the low hangar headroom. In one respect, however, all six ships were well able to accept new and heavier aircraft as their armoured decks were tremendously strong and well able to accept the increased loads. Nevertheless, the overriding constraint was financial and the plain fact was that post-war Britain was in no condition to fund major refits as well as build new carriers and produce the numbers of aircraft required to equip them. Consequently, some hard decisions had to be made and the policy of investing in the new ships was adopted at the expense of modernising some of the existing ships.

While the Royal Navy struggled to maintain a nucleus of large fleet carriers, it was able to maintain an effective overseas deployment of naval air power by the use of the Colossus class light fleet carriers. Ten of these were completed and although two were modified as aircraft repair ships and could not operate aircraft, the remaining eight rendered sterling service and several were transferred to foreign navies where they remained in service for several decades. In particular the Royal Navy maintained at least one light fleet carrier in the Far East throughout the Korean War. Those involved were HMS *Triumph*, HMS *Theseus*, HMS *Glory* and HMS *Ocean*, all of these operating air groups consisting of Sea Furies and Fireflies except for *Triumph*, which was the first carrier off Korea and at that time was equipped with Seafire 47s and Firefly FR.1s. In Royal Navy service these light fleet carriers were little modified for normal service, a tribute to the soundness of the basic design. However, at one time or another some were involved in some interesting experiences. HMS *Vengeance* made a cruise to Arctic waters in early 1949 to carry out trials of aircraft operation in extreme cold weather conditions. Aircraft embarked for a six-week period included a Sea Vampire jet fighter and a Dragonfly helicopter. Also in 1949 HMS *Warrior* was modified to conduct trials of the bizarre rubber deck experiment and in 1951 HMS *Triumph* was the first carrier to conduct trials of the angled deck concept. Both of these events are described in detail in Chapter 5.

The Colossus class displaced 15,690 tons (standard) and the normal aircraft complement was thirty-five, although this could be increased to forty-five under war conditions. The flight deck measured 695 feet by 80 feet, only some 60 feet shorter than the Illustrious class (740 to 760 feet) but 15 to 20 feet narrower, which did put a wingspan limitation on the type of aircraft that could be operated. This wasn't significant in respect of standard post-war types such as the Seafire, Sea Fury and Firefly but precluded larger aircraft such as the Firebrand and Sea Hornet. In an effort to speed production, the design of these ships was based on mercantile principles and propulsion was provided by two sets of standard destroyer machinery to give a speed of 25 knots, slower than the fleet carriers but sufficient for most operational purposes. As already mentioned several found their way into service with other navies. The first to go was *Colossus* herself, which was transferred to France in 1946 and renamed *Arromanches*. In 1948 HMS *Venerable* was transferred to the Royal Netherlands Navy as HMNS *Karel Doorman*, later being resold to Argentina in 1958 where as the *25 de Mayo* she posed a threat to British forces during the 1982 Falklands War, although no operational missions were flown due to catapult problems. HMS *Warrior* was loaned to the RCN from 1946 to 1948 and after further Royal Navy service was sold to Argentina in 1958. Finally, HMS *Vengeance* was sold to Brazil in 1957 and remained in service for several decades.

In addition to the ten Colossus class, a further six Majestic class were laid down in late 1943 and these were basically a repeat of the basic design except that the flight deck was strengthened and larger lifts (54 feet by 34 feet) were fitted to permit the operation of larger and heavier aircraft. Although all were launched in 1944/5, none had been completed by the end of the War and they were laid up while decisions were made as to their future. In fact, none were destined to serve with the Royal Navy. HMS *Magnificent* was completed in 1948 and transferred on loan to the RCN until 1957, after which she was laid up and scrapped. She had been completed to the original design with an axial flight deck. Her sister ship, HMS *Powerful*, was sold to Canada but was completed to a substantially revised design with an angled deck to enable jet operations before being delivered in 1952 as HMCS *Bonaventure*. A similar arrangement occurred with the Royal Australian Navy, which took over HMS *Terrible* as HMAS *Sydney* in 1948, completed with a conventional axial flight deck. She subsequently saw service in the Korean War and was not finally retired until 1973. The RAN also took delivery of HMS *Majestic*, renamed as HMAS *Melbourne* in 1955, in this case completed to a totally revised design incorporating an angled deck and other equipment to allow the operation of Sea Venom jet fighters and Gannet ASW aircraft. The last of the class to be completed was HMS *Hercules*, which was sold to India in 1957 and commissioned as INS *Vikrant* in 1961, by which time she also had an angled deck, steam catapult and mirror landing sight. The sixth ship of the Majestic class, HMS *Leviathan*, was never completed and was eventually scrapped in 1968, although her boilers and turbines were sold to the Netherlands for use in a refit of the *Karel Doorman*.

One of the reasons why the Majestic class carriers were not completed for the Royal Navy was that, apart from the fact that the existing Colossus class were deemed sufficient for the immediate post-war era needs, the service also had four Hermes class light fleet carriers under construction. With a standard displacement of over 20,000 tons and a flight deck measuring 733 feet by 103 feet, these ships were originally designed to carry fifty aircraft and were a substantial improvement on the Colossus and Majestic classes. Originally, eight ships were projected but in the event only four were completed and the first of these, HMS *Centaur*, did not commission until 1953. Consequently these ships are described in more detail in Chapter 5.

While the Royal Navy effectively lived from hand to mouth in the early post-war years and delayed the completion of newer carriers, the US Navy was in a much more fortunate position. Its numerous Essex class carriers were already in commission and the three Midway class ships were all completed and retained in service in the late 1940s. Of the twenty-four Essex class, most were already in service in September 1945 and the final six built to the standard design were all completed by the end of 1946. One more ship, the USS *Oriskany*, was launched in October 1945 but her completion was delayed to incorporate numerous modifications and differed in several aspects from her sister ships when she entered service in 1950. Nevertheless, even the US Navy did not require all of these in peacetime commission and only four of the newer ships were retained on active service (*Boxer*, *Leyte*, *Valley Forge* and *Philippine Sea*), the remainder being laid up in reserve. Although superficially similar in appearance, the standard Essex class carriers were in fact built as two separate sub groups, the first ten ships being completed to the original design with an almost straight stem and a single quadruple 40 mm gun on the bows below the flight deck overhang. The remaining thirteen had a lengthened bow and two quadruple 40 mm mountings were carried side by side, giving an easy recognition point. Interestingly, the Essex class originally perpetuated the US Navy practice of equipping carriers with arrester wires on the forward flight deck so that aircraft could be landed while the ship went astern. The first five ships were also fitted with a hangar deck catapult capable of launching an aircraft on the beam. Both of these features were deleted during the course of the War and an additional deck catapult was installed in the bows. The Essex class carriers were then normally equipped with two H4 catapults, which were capable of launching all the various wartime aircraft types but struggled to cope with some of the larger types embarked after 1945.

Inevitably these ships were modernised over the years but in the case of the Essex class the process was complex and varied. Initially there were ambitious plans to convert them to operate pilotless aircraft and guided missiles but by the end of 1946 the emphasis was on carrying an air group of heavy attack aircraft and jet fighters. This reflected the US Navy's perceived role of supporting operations against land targets as there no longer existed any potential naval power that would provide any credible sea-based opposition (except for submarine attack and ASW was taken very seriously, although this resulted in differing lines of development described in Chapter 9). The project to convert the Essex class carriers to the new role was termed SCB-27A (the acronym SCB deriving from Ships Characteristic Board) and incorporated some drastic changes. These included eliminating the island superstructure to provide a flush flight deck and the full load displacement would have risen to in excess of 40,000 tons. However, even the US Navy with all its resources could not afford such an ambitious project and the whole thing was scaled down to the minimum necessary to operate prospective jet fighters and the heaviest attack aircraft without major structural alterations.

The conversion programme commenced in 1948 and over the next five years eight ships were modified to SCB-27A standard. The immediate objective of this programme was to equip the ships to operate aircraft up to 18 tons (approximately 40,000 lb),

this being sufficient to deal with new heavy piston-engined
types such as the Skyraider, Mauler and Guardian, and with
the early generations of jet fighters, few of which exceeded
25,000 lb. The most obvious change was the installation of new
H8 catapults capable of launching such aircraft and these were
considerably longer than the H4s that they replaced. These
were still conventional hydraulic catapults in which pressure
drove a piston through a bore, its motion being magnified by a
system of cables and pulleys to increase the actual throw of the
catapult. In the H8 this system reached its practical limit and
further improvements to cope with the new heavy attack aircraft would require a completely new approach. To cope with
larger and heavier aircraft the SCB-27A ships had strengthened
flight decks and larger lifts of greater capacity. The deck edge
lift could now accommodate aircraft up 30,000 lb and the flight
deck lifts were rated at 40,000 lb, while the flight deck could
cope with parking and take-offs by aircraft of this weight, but
could only accept aircraft up 30,000 lb landing weight. Not so
obvious were a number of important improvements to enable
the operation of jet aircraft whose arrival was having a considerable impact on the ships themselves. For example, jet aircraft
used considerably more fuel than the piston-engined types they
replaced. The F2H Banshee entering service in 1949 carried
over 16,000 lb of fuel compared with a typical 2,400 lb uplift by
an F6F Hellcat. The Essex class carriers were originally designed
to carry 209,000 US gallons of aviation fuel and even the later
Midway class only carried 330,000 US gallons when first commissioned. In the SCB-27A modernisation this capacity was
increased by approximately 30 per cent, a process partly made
easier by the fact that the standard kerosene-based jet fuels
were much less volatile than the gasoline used in piston engines.
    Another feature required for jet operations was the fitting
of retractable blast screens behind the catapult launch points
to protect other aircraft ranged on deck. Here, the coming of jet
aircraft again had a major impact on deck operations. In the
case of a deck load of wartime piston-engined fighters and
strike aircraft, the favoured method of launch was to carry out
free rolling take-offs along the deck. In some cases the first few
aircraft of a strike force would be catapulted off until enough of

the deck was available for the remainder to carry out rolling take-offs. Jet aircraft, however, required much longer take-off runs and consequently fewer aircraft could be ranged. In fact, later versions of the Banshee and Panther could not take off in the length of deck available, even with a strong wind over the deck. In these circumstances a catapult launch was essential and deck-operating routines had to be adapted to reflect this.

Landing aircraft also required changes. The arrester gear needed be considerably uprated to deal with heavier aircraft landing at higher speeds and even then a longer roll out was required to absorb the energy of the landing. This further reduced the forward area available for a deck park. The SCB-27A ships had thirteen arrester wires instead of the original eleven and had to be fitted with several types of barriers to catch aircraft that did not engage the wires. Traditionally this took the form of steel cables stretched across the deck amidships whose function was to stop landing aircraft running into the forward deck park. In piston-engined aircraft this was an acceptable method as the pilot sat well back, protected usually by a large radial engine and propeller. Although the aircraft was quite often seriously damaged, the pilot usually survived and damage to other aircraft was prevented. However, it was soon realised that such barriers were not suitable for jet operations, the taut steel wires being potentially lethal to the pilots who now sat well forward in the nose of the aircraft with little to protect them. Eventually a new type of barrier utilising a series of vertical nylon webbing straps suspended from a cable strung high across the deck was introduced (this was a British design that had been introduced aboard HMS *Eagle* when she commissioned). With this type of barrier the nose of an overrunning aircraft would pass below the cable and through the loosely hung nylon straps, which would then catch the wings and bring the aircraft to a halt. This not only offered better protection to the pilot but generally resulted in less damage to the aircraft.

Another aviation-related factor that impinged on the ships was the greater weight and variety of ordnance that the newer aircraft could carry. In World War II a load of two 1,000 lb bombs was as much as many aircraft could carry, even dedicated attack aircraft such as the Avenger and Helldiver. The post-war generation of aircraft such as the Skyraider and Mauler could carry up to 8,000 lb and weapons available included not only conventional bombs but rocket projectiles of varying sizes (including the 11.75 inch Tiny Tim) and later came air-to-air and air-to-surface guided missiles. All these required safe stowage in magazines below deck, and lifts and hoists to bring them up to deck level. A whole host of specialised handling equipment was then needed to move the ordnance around the flight deck and lift it into position for arming the aircraft.

Wartime experience showed the need for adequate aircrew rest and briefing facilities and these required easy access to the flight deck, a passage made harder by the increasing accoutrements collected by crews, including g-suits, parachutes, 'bone domes' and other specialised flying equipment. In the Essex SCB-27A modernisation these issues were addressed and a moving escalator installed to enable to crews to quickly reach the flight deck with ease. Other changes included a complete restructuring of the island superstructure with the single funnel now faired in behind the bridge and a taller mast stepped to allow optimum installation of a new suite of radars including SPS-6 search, SPS-8 height finder and a CCA radar. All these changes inevitably had a weight penalty, which affected the performance and stability of the ship. To counteract this, beam was increased, waterline armour was removed and all the deck-mounted twin 5 inch gun mounts were removed, although some were replaced by additional single open mounts below the edge of the flight deck, while all the 40 mm and 20 mm guns were removed and replaced by a battery of radar-directed 3 inch/50 cal guns. The final armament was eight 5 inch/38 cal guns and twenty-eight 3 inch/50 cal in twin mountings.

The SCB-27A modifications were extended to eight ships in all (*Essex, Yorktown, Hornet, Randolph, Wasp, Bennington, Kearsarge* and *Lake Champlain*). In addition, work on the twenty-fourth and last of the class, USS *Oriskany*, was delayed so that she could be completed to SCB-27A standard.

During the period under review in this chapter the three Midway class carriers entered service and initially few modifications were required as due to their size there were initially

Eugene Ely made the historic first landing aboard a ship flying a Curtis pusher biplane on 18 January 1911 using a temporary deck erected on the stern of the cruiser USS *Pennsylvania*. The same pilot had made the first take off from a ship in November 1910. These early flights represented the start of shipboard naval aviation. (*US National Archives*)

The USS *Saratoga* was one of two large carriers which were converted from battle cruisers under construction at the end of the First World War. This photo was taken in 1938 and shows the ship's full air group ranged on deck. Although almost all of the aircraft are biplanes, a small number of TBD Devastator monoplane dive bombers can also be seen. (*US National Archives*)

The Griffon engined Supermarine Seafire Mk.XVII was the major post-war variant of this famous fighter and entered service in September 1945. Subsequently it served with various RNVR and training squadrons until 1954. (*Fleet Air Arm Museum*)

The Supermarine Seafang sought to improve performance by adopting a new profile laminar flow wing. Along with the RAF's Spiteful, these were the fastest British piston-engined fighters ever built but did not enter operational service. The photo shows one of ten Seafire F.Mk.31s which were flown for evaluation purposes. (*Fleet Air Arm Museum*)

An early production Hawker Sea Fury F.Mk.10 takes off from the light fleet carrier HMS *Ocean* in 945. The Sea Fury was probably the best all round British piston engined fighter but was too late to ee service in the Second World War, although it played a key role in Royal Navy support perations off Korea. (*Fleet Air Arm Museum*)

A prototype de Havilland Sea Hornet F.Mk.20 single seat fighter gets the cut signal from the atsman as it carries out deck landing trials in June 1946 aboard HMS *Ocean*. This version equipped nly one frontline squadron as did the two seat N.F.21 which subsequently entered service in 1949. *Fleet Air Arm Museum*)

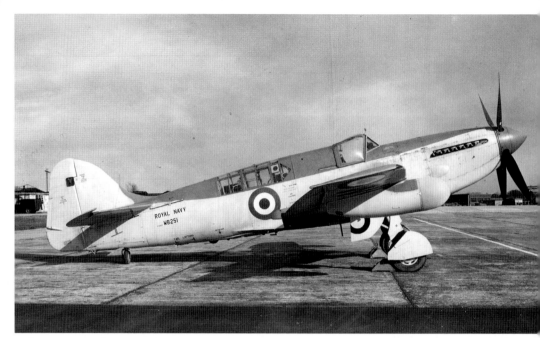

The Fairey Firefly was an important element of Royal Navy carrier air strength in the post-war decade, being produced in several different versions based on the Firefly FR.4 which had flown in 1944. Photo shows a Firefly AS.5 anti-submarine variant which was equipped with ASV radar and carried depth charges and sonar-bouys. (*Fleet Air Arm Museum*)

A Hawker Sea Fury comes to rest after engaging the crash barrier aboard HMS *Triumph* in 1951. This was a not uncommon event prior to the introduction of the angled deck. (*Fleet Air Arm Museum*)

2935

After a protracted period of development, the Blackburn Firebrand eventually entered service in September 1945, replacing the Barracuda as the Royal Navy's sole torpedo bomber. This T.F.4 shown here displays the much larger fin and rudder introduced with this variant to improve directional stability.

(Fleet Air Arm Museum)

The Blackburn Firecrest was intended as a replacement for the Firebrand, but the performance benefits were insufficient to warrant production. (*Fleet Air Arm Museum*)

The Fairey Spearfish flew in July 1945 and was the first British naval bomber to feature an internal bomb bay. Despite its promise, production orders were cancelled at the end of the Second World War and only four aircraft were flown. (*Fleet Air Arm Museum*)

The Royal Navy's last propeller driven strike aircraft was the Westland Wyvern. After a lengthy development programme involving no less than three different powerplants, the resulting Wyvern 4 entered service in 1953. This example landing aboard HMS *Eagle* belongs to 827 NAS. (*Fleet Air Arm Museum*)

The F8F-2 Bearcat was the last in a line of famous Grumman piston engined naval fighters. Entering service just too late to see action in the Pacific War, its prime role was as a fast climbing interceptor and as such was quickly rendered obsolete by the new generation of jet fighters. (*US National Archives*)

Essex class carrier USS *Kearsarge* (CV.33) shown while operating in the Mediterranean in 1948. The aircraft ranged on deck include Bearcats, Avengers and

Blackburn Firebrands of 813 NAS warm up aboard HMS *Indomitable*. In the background are the battleship HMS *Vanguard* and the carrier HMS *Implacable*. (*Fleet Air Arm Museum*)

The swept wing Supermarine Type 510 was a development of the straight winged Attacker. Unlike its predecessor, the Type 510 did not see operational service but it was the first swept wing aircraft to land aboard an aircraft carrier, this event occurring on 8 November 1950. (*Fleet Air Arm Museum*)

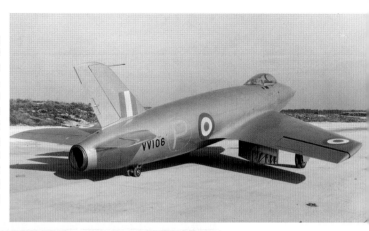

The Supermarine Attacker was the Royal Navy's first operational jet fighter, entering squadron service in 1950. The wing was almost identical to that first tried out on the propeller driven Seafang in 1944. (*Fleet Air Arm Museum*)

The first deck landing by a jet aircraft occurred on 3 December 1945, the aircraft being a de Havilland Vampire flown by Lieutenant Commander Eric Brown RN shown here at the moment of catching the wires aboard the light fleet carrier HMS *Ocean*. (*Fleet Air Arm Museum*)

A pair of Sea Vampire F.20s aboard HMS *Theseus*. Although never deployed with front line squadrons, the Sea Vampires enabled the Royal Navy to gain considerable experience of operating the new jets at sea. (*Fleet Air Arm Museum*)

The United States utilised several aircraft as flying testbeds for the new jet engines. Shown here is a Goodyear built Corsair fitted with an experimental Westinghouse WE-19X turbojet under the fuselage centre section. (*US National Archives*)

The Curtiss XF15C-1 was an unsuccessful attempt to produce a mixed power fighter. This had a 2,100 hp radial piston engine in the nose and a 2,700 lb thrust turbojet in the rear fuselage. Only three prototypes were flown. (*US National Archives*)

Vought's first foray into jet fighters was the F6U-1 Pirate which flew in prototype form in October 1946. Although put into limited production, it was only used for trials and training and the US Navy preferred other aircraft such as the North American Fury and Grumman Panther for operational squadrons. (*US National Archives*)

The US Navy's first jet fighter to be deployed t sea as part of a regular carrier air group was th straight winged North American FJ-1 Fury. Th aircraft shown here overflying San Francisc in 1949 belong to an Oakland based Reserve Squadron. (*US National Archives*)

A North American FJ-1 Fury of VF-5A makes a rolling take off from the deck of the Essex class carrier USS *Boxer* (CV.21) in March 1948. (*US National Archives*)

pair of McDonnell F2H-2 Banshees overfly ships of Task Force 77 off Korea in July 1953. The twin gined Banshee was one of the most successful of the early American naval jet fighters and served ith the US Navy for over a decade from 1948 onwards. In addition, thirty-five were transferred to e RCN where they remained operational until 1962. (*US National Archives*)

Despite advances in technology, naval aviation was always a dangerous occupation. This Banshee hit the round down when attempting to land on the USS *Oriskany* in 1954. The aircraft broke up and burst into flames but, miraculously, the nose section rolled clear and the pilot suffered only minor injuries! (*US National Archives*)

A pilot climbs into the cockpit of his F9F-2 Panther aboard USS *Boxer* in October 1949. By the outbreak of the Korean War in 1950, the US Navy had no less than six Panther squadrons at sea. (*US National Archives*)

A dramatic shot showing a Supermarine Attacker of 803 NAS being prepared for a catapult launch aboard HMS *Eagle*. Note the under fuselage auxiliary fuel tank. (*Fleet Air Arm Museum*)

The most significant of the first generation Royal Navy jet fighters was the Hawker Sea Hawk which equipped no less than thirteen frontline squadrons between 1953 and 1960. This view shows the first prototype (VP413) carrying out deck landing trials aboard HMS *Illustrious* on July 1949. (*Fleet Air Arm Museum*)

A de Havilland Sea Vampire picks up the single arrester wire during trials of the flexible deck concept aboard HMS *Warrior* early in 1949. Although the trials were a great success, the concept was not adopted for operational use. (*Fleet Air Arm Museum*)

A Blackburn Firebrand of 827 NAS makes a sorry sight after engaging the crash barrier aboard HMS *Eagle*. Efforts to avoid this type of scenario eventually led to the adoption of the angled deck concept. (*Fleet Air Arm Museum*)

The US Navy were quick to modify the Essex class carrier USS *Antietam* in 1953 to test the angled deck concept. The ship is shown here carrying out trials with a variety of then current aircraft on deck including Panthers, Cougars, Banshees and a single FJ-2 swept wing Fury. (*US National Archives*)

few problems encountered in handling the first generations of naval jet aircraft. All three had their flight decks strengthened in 1947–8 in order to operate the new AJ-1 Savage nuclear bomber. Although *Midway* and *Franklin D Roosevelt* were completed in 1945 with the designed main armament of eighteen single 5 inch/54 cal guns, the *Coral Sea* did not commission until 1947 and carried only fourteen 5 inch guns and no 40 mm light AA weapons. However, in 1948 the armament of all three was standardised as fourteen single 5 inch/54 cal and twenty twin 3 inch/50 cal mountings with any existing 40 mm and 20 mm guns being removed. More extensive changes were made in the 1950s and these are described in Chapter 5.

4

THE COMING OF THE JETS

As far as the victorious allies were concerned in late 1945, the jet engine was the invention of the British pioneer, Sir Frank Whittle. Certainly, it was British pioneering work that resulted in the RAF being able to introduce the twin jet Gloster Meteor in 1944 and provided the basis for the American aircraft industry to enter the jet age during the war. However, although they started later, German engineers made rapid progress between 1935 and 1939, flying the world's first turbojet-powered aircraft (Heinkel He.178) on 27 August 1939. Subsequently, the twin-engined He.280 jet fighter flew in April 1941. By the end of the European War in May 1945, the jet-powered twin-engined Me.262 fighter and twin/four-engined Ar.234 bomber were both in frontline service, while numerous prototype jet designs were flying or under development. Without exception these were powered by axial flow turbojets of advanced design offering a relatively small frontal area, although due to problems related to rushed development and a shortage of vital materials, they were notoriously unreliable.

Meanwhile, in Britain, Flt Lt Frank Whittle had started work on developing his own ideas of jet propulsion during the 1920s and registered his first patent in January 1930 for a jet engine that had all the basic components utilised in subsequent designs. It consisted of a single-stage compressor that fed air at high pressure into a combustion chamber where it was heated by burning fuel and exhausted through a turbine, which in turn drove the compressor. The whole cycle was continuous and had the great virtue of being mechanically very simple, although the problem of manufacturing components capable of operating at high rpm under extreme temperatures was to be the element that determined the rate at which jet engine development proceeded. Initially this was at an agonisingly slow rate as there was no official support forthcoming and Whittle was forced to rely on private contributions to fund his work. Consequently it was not until April 1937 that he was able to run his first complete jet engine, which had been built to his specifications by British Thompson Houston (BTH). By that time the German Hans von Ohain had already run the HeS.1 axial flow jet engine and in the following year Junkers was bench running an axial flow engine of its own design.

In Britain, official interest was awakened by the outbreak of World War II and an injection of funding allowed Whittle to form Power Jets Ltd and produce the Whittle W.1, which delivered 850 lb thrust. Like most of the Whittle designs, this featured a double-sided centrifugal compressor in which a single large fan drew in air. This was compressed by a combination of the aerodynamic properties of the intake ducting and the centrifugal energy imparted to the air, which was thrown outwards to the edge of the fan disc before being led into the combustion chambers. This layout had the virtue of simplicity, the fan being cast in a single piece, but meant that such engines had a relatively large frontal area compared with the axial flow design in which the air flows straight through a series of compressor fans. This arrangement was much more complex in engineering terms but allowed much slimmer powerplants and ultimately became the standard for almost all jet engines. However, in the early days British development centred on the centrifugal flow jet engine and Whittle's Power Jets company became primarily a research and development establishment with ideas being passed on to established industrial concerns who subsequently designed and built their own engines. Initially this was the Rover company, which built and tested the W.2B on Whittle principles and by September 1942 this was producing 1,700 lb thrust. At this point Rover was forced to withdraw from jet engine development and its work was taken over by Rolls-Royce, who developed the W.2B into the Welland turbojet that powered initial production versions of Britain's first jet fighter, the Gloster Meteor. However, the prototype Meteor that first flew on 5 March 1943 was actually powered by two de Havilland H-1 Goblin turbojets delivering

1,500 lb thrust as these were the first flight-capable engines available. De Havilland had become involved in gas turbine development in 1941 when it began work on a new single-engined jet fighter. This was initially and rather awkwardly known as the Spider Crab, a name perhaps inspired by its twin boom layout, which eventually became the Vampire and flew in prototype form on 26 September 1943. Major Halford, de Havilland's engine designer, produced an engine that was simpler than the Whittle designs and featured a single-sided compressor linked to a straight-through combustion system (early Whittle engines had a complex reverse-flow system in an attempt to reduce overall length).

These two lines of development gave birth to a range of jet engines that gave the British aero engine industry a commanding lead in jet propulsion in the immediate post-war years. On the Rolls-Royce side, the original Welland set a tradition for naming their engines after rivers. It was superseded by the more powerful Derwent, which was initially rated at 2,000 lb thrust but was ultimately capable of delivering over 4,000 lb in late production versions. A scaled-up version became the Nene and this was first run in October 1944. It was developed to give over 5,000 lb thrust and was to play an important role in the evolution of jet fighters for the Royal Navy. The de Havilland engine range started with the Goblin. This eventually produced around 3,500 lb thrust and was followed by the Ghost, which gave over 5,000 lb in its final versions.

For most of the period covered by this book, the centrifugal flow engines like the Nene and Goblin powered the majority of frontline fighters for both the RAF and Royal Navy. However, axial flow engines had not been neglected and in particular Metropolitan-Vickers had started work on this type of engine as early as 1940. By 1945 the company had test flown the 1,800 lb thrust F.2 and 3,500 lb thrust Beryl. The latter engine powered the unique Saunders Roe S.R.A.1 flying boat fighters, which are described later. A scaled-up version was named the Sapphire but development of this axial flow turbojet was handed over to Armstrong Siddeley in 1948 and ultimately it gave over 10,000 lb thrust. Rolls-Royce also commenced work on an axial flow turbojet that became the Avon. This first ran in 1945 and subsequently was the main powerplant for a whole range of British aircraft, including the Hunter, Javelin and Canberra as well as later versions of the Comet civil airliner. Thrust rating started at 6,500 lb thrust and eventually rose to over 16,000 lb thrust with afterburning.

The development of the jet engine was closely monitored by the Royal Navy and it was particularly interested in the Vampire, which, compared with the twin-engine Meteor, was regarded as more suitable for carrier operations due to its smaller size and simpler design. By mid 1945 orders for 300 Vampires had been placed for the RAF and the first production F.1 flew in April 1945. The second prototype Vampire (LZ551) was then made available for modifications so that carrier deck landing trials could be carried out. A V-frame arrester hook was fitted below the jet exhaust pipe, flap area was increased by 40 per cent and long stroke undercarriage oleos were fitted. Subsequently an uprated Goblin 2 engine was installed and a new clear-view canopy provided, and in this form the aircraft was delivered to a Royal Navy trials unit at Ford for carrier trials. After a series of dummy deck landings at the airfield, the aircraft was flown out to meet the light fleet carrier HMS *Ocean* steaming in the English Channel on 3 December 1945. Piloted by Lt Cdr Eric 'Winkle' Brown, then commanding officer of the RAE Aerodynamics Flight, the Vampire made a successful landing off the first approach. In fact, the Vampire was ideal for deck landing, being rock steady on approach and giving the pilot an unprecedented clear view of the deck at all times – a stark contrast with the then current generation of piston-engined naval fighters such as the Corsair, Seafire and Sea Fury in which a long nose or big radial engine effectively shielded great swathes of the deck from view. Take-offs also proved to be relatively straightforward, the Vampire being already airborne as it passed the bridge on the ship's island superstructure. On this epoch-making day, the Vampire made four landings and takes-offs but had to be flown ashore after the last landing due to flap damage. The flaps were subsequently reduced in size and no further problems were encountered.

The Royal Navy had taken a significant step forward in becoming the first to fly a jet-powered aircraft from the deck of a carrier but, as was so often the case, it then failed to capitalise on a pioneering effort and the lead was soon to slip to the Americans. Following the success of the initial trials, a second Vampire was converted and the two aircraft became designated Sea Vampire F.10. Two further prototypes were ordered of what was to become the Sea Vampire F.20. This version was based on the RAF's Vampire FB.Mk.5, which featured a strengthened airframe and had clipped square wingtips, reducing span by 2 feet to 38 feet. An A-frame arrester hook was fitted and repositioned to sit atop the jet pipe when stowed. Eighteen production Vampire F.20s were ordered and these were delivered between October 1948 and June 1949. These Vampires were intended only to allow naval pilots to familiarise themselves with jet operations and were flown by shore-based trials and training squadrons including Nos 700, 702 and 787. Despite not having folding wings, occasional carrier detachments were flown including Exercise *Sunrise*, in which over 200 deck landings were made, and an Arctic cruise by HMS *Vengeance* in 1950 in which the operation of a number of new naval aircraft including the Vampire were tested in extreme weather conditions. One of the prototype F.20s gained the distinction of being the first jet aircraft to be catapulted from a carrier deck and this event occurred on 15 July 1948. Sea Vampires continued to contribute to many aspects of carrier development and were involved in trials of the new mirror deck landing sight while three more Vampires were modified to F.21 standard for trials connected with flexible rubber decks (described in Chapter 5). Thus although the Sea Vampire never equipped an operational squadron, it provided invaluable experience to a whole range of naval pilots, airmen, sailors and engineers and provided the basis on which the Fleet Air Arm could introduce new frontline naval jet aircraft in the 1950s.

While Britain and the Royal Navy were experimenting with naval jet operations, across the Atlantic the Americans were quick to see the potential of the jet engine and even quicker to catch up and overtake their mentors. Even prior to 1941, some progress had been made and both Lockheed and Northrop had proposed aircraft designs to be powered by jet reaction engines. Experiments had been conducted by NACA using the Italian

Caproni-Campini system in which a piston engine drove an axial flow ducted fan, the airflow then being boosted by an after-burning system. Although thrust ratings of over 2,000 lb thrust were obtained in ground tests, the programme was abandoned in 1943 as it became obvious that fuel consumption was much too high for flight purposes. American jet progress received a tremendous boost when General Arnold, then C-in-C of the USAAF, visited Britain and saw for himself the progress being made. With his backing, two Whittle engines and sets of draw-ings for a W./2B were sent to America where the General Electric Company was entrusted with their development. At the same time, in September 1941, the Bell company was contracted to design and build a new jet fighter under the designation XP-59A Airacomet (this was something of a security smokescreen as the XP-59 designation had already been applied to a now defunct Bell twin piston-engined design). Three XP-59 prototypes were ordered and the first was completed remarkably quickly, making its first flight on 2 October 1942 – almost six months ahead of the British Gloster Meteor! It was powered by two GE I-A turbojets rated at 1,400 lb thrust and was a relatively simple design with a mid-mounted straight tapered wing and the two engines faired into the underside of the wing roots. Unfortunately this arrangement caused severe aerodynamic interference between the engines and airframe with a result that performance was disappointing with a top speed of only just over 400 mph. The original armament consisted of two nose-mounted 37 mm cannon and this was fitted to most of the thirteen YP-59As ordered in 1942, which were powered by improved 1,650 lb thrust General Electric I-16 engines. Twenty production P-59As were delivered, these carrying an armament of one 37 mm cannon and three 0.5 inch machine-guns but the Airacomet proved to be a very poor gun platform due to inherent directional instability and orders for eighty P-59Bs were cut back after only thirty had been delivered, production ending in May 1945. The P-59 Airacomet provided US pilots with their first taste of jet operations and was utilised mainly as a fighter trainer in preparation for the introduction of better aircraft.

Although most P-59s went to the USAAF, two YP-59As were delivered to the US Navy as the Bell XF2L-1. These were the eighth and ninth aircraft off the production line and were delivered in November 1943 (BuNo 63960 and 63961). Subsequently the US Navy also received three P-59Bs in 1945. These aircraft gave the US Navy its first experience of jet operations but they remained strictly land-based and were never embarked on carriers. In the meantime the US Navy had looked at the idea of fitting a jet engine as a booster to an existing design. The aircraft selected was the Douglas BTD-1 Destroyer, which underwent testing in 1944. Two production aircraft were modified during construction by installing a 1,500 lb thrust Westinghouse WE-19X turbojet in the rear fuselage, which was canted up at 10 degrees and was fed via a dorsal intake behind the cockpit with the jet pipe exhausting below the rear fuselage. A jet fuel tank was fitted in the bomb bay. The first flight took place at Los Angeles Municipal Airport in March 1944 but subsequent tests showed that the expected 50 mph boost in top speed could not be achieved as the turbojet performance fell off drastically at speeds above 200 mph due to problems associated with intake airflow. In view of this, as well as the emergence of other more successful aircraft, the programme was ended shortly after the end of the war. The WE-19X had been designed by Westinghouse specifically to a US requirement for a booster jet engine and had begun ground testing in March 1943 while a second engine was fitted under a Chance Vought F4U Corsair for flight trials, these commencing on 21 January 1944. Subsequent development resulted in the J-30, which was ordered into large-scale production in early 1945, although the order was sub-contracted to Pratt & Whitney.

The Whittle engines were not the only jet designs to be imported from Britain and examples of the de Havilland H-1B Halford centrifugal flow engine were also obtained. These were initially destined for the Lockheed P-80 Shooting Star single-seat fighter ordered for the USAAF. Early versions of this engine developed around 2,700 lb thrust and the US Navy obtained some for evaluation purposes. For flight trials a Grumman Avenger was selected as the test bed as its capacious bomb bay provided space to install the jet engine and its ancillary systems. The particular aircraft was one of two Grumman-built XTBF-3 prototypes produced to test the installation of a 1,900 hp

Wright R-2600-20 radial piston engine, which would eventually replace the original 1,600 hp R-2600-8 used in the earlier TBF/ TBM-1 series. The jet engine was mounted beneath the belly of the aircraft with twin intakes positioned either side of the lower fuselage in line with the rear of the piston engine cowling. The single exhaust was on the under fuselage centreline, just aft of the wing trailing edge. To compensate for the extra weight, all armament was removed and in place of the dorsal gun turret was a partially glazed fairing, providing accommodation for a flight test observer. Conversion work was carried in mid 1944 and ground testing began in November of that year, with flight tests commencing the following month. This was an historic occasion bearing in mind Grumman's subsequent success as a builder of naval jet fighters and much useful data was obtained. The experience with a British engine was also to bear fruit subsequently but in the early post-war years Grumman initially lost out to other manufacturers.

Perhaps Grumman's greatest rival was to be the fledgling McDonnell company, which had only been formed in 1939 but had gained a reputation for innovative engineering and had already carried out some preliminary studies on the application of jet propulsion to combat aircraft. Consequently, when the US Navy began to take an active interest in the idea of carrier-based jet fighters, McDonnell were well placed to bid for the contract. Another factor was that other major manufacturers such as Douglas and Grumman were fully occupied with projects of their own and had little spare capacity. When McDonnell submitted their proposals in late 1942, they were quickly rewarded with a Letter of Intent authorising them to proceed with the design and construction of two prototype jet fighters under the designation XFD-1 and the name Phantom was subsequently adopted. The specification called for a fleet air defence fighter capable of carrying out combat air patrols over its parent carrier at an altitude of 15,000 feet. This was a remarkably loose specification as other important items such as performance and armament were not set out.

The design team led by Kendall Perkins adopted a principle of keeping the airframe as simple as possible so that a low-set straight wing was adopted, although a tricycle undercarriage was selected in order to keep jet blast away from the wooden decks of the American carriers. At this stage jet engines were very much in their infancy and it was thought that they could be produced in any required size, overall dimensions being scaled up or down to meet different requirements. Consequently, early XFD-1 design proposals included no fewer than six 300 lb thrust engines buried in the wings – each engine having a diameter of only 9.5 inches! Other configurations investigated included eight- and ten-engined versions but eventually it was found that a more conventional twin-engined layout offered the best performance, saving both weight and complexity. The selected powerplant was the Westinghouse WE-19XB-2B axial flow turbojet, which had a thrust rating of 1,600 lb and an external diameter of 19 inches.

Construction of XFD prototypes began in January 1944 and the airframes were quickly completed but delays with engine development meant that by October 1944 only a single airworthy example had been delivered. Nevertheless this was installed and following ground runs the prototype undertook high-speed taxiing tests, actually becoming airborne for a brief moment on 2 January 1945. Eventually a second engine was delivered and installed and the first full-scale flight took place on 26 January 1945. Flight trials showed that the XFD-1 Phantom was a pleasant aircraft to fly with only minor problems such as excessive aileron system friction and some directional stability issues, all of which were amenable to rectification. The major factor that delayed the test and development programme was the unreliability of the early engines, while the first prototype was lost in a crash caused by aileron failure in November 1945. Nevertheless the US Navy was confident enough to order 100 FD-1s, although this was later cut back to only sixty aircraft, which was still sufficient for test and evaluation purposes and for two operational squadrons to be formed. After some delay due to a lack of engines, these were all delivered between January 1947 and May 1948 under the designation FH-1 – McDonnell had been allocated the letter H to avoid confusion with Douglas aircraft (D).

The standard FH-1 Phantom was powered by two 1,600 lb thrust Westinghouse J30-WE-20 turbojets, which gave a top speed of 479 mph at sea level, an initial rate of climb of 4,950 feet/min and a range of 540 miles. The service ceiling was 41,000 feet. The armament comprised four 0.5 inch machine-guns mounted in the nose and there was no provision for external ordnance. The production aircraft could be distinguished by an enlarged tail fin with a squared off top and a slightly lengthened fuselage, which allowed an increase in fuel capacity. The aircraft's performance figures were generally better than contemporary piston-engined fighters except in respect of range and endurance, a common shortfall with the early jets and one of the reasons why the Royal Navy had not used the Sea Vampire to equip an operational squadron. However, the US Navy had no such reservations and as soon as the aircraft were available the first jet fighter squadron, VF-17A, was formed at the end of 1947. In the meantime one of the prototype XFD-1s had become the first US jet to land aboard an aircraft carrier, this significant event occurring on 21 July 1946 – seven months after Lt Cdr Brown had landed aboard HMS *Ocean*. The US pilot was Cdr Jim Davidson and the trials were conducted aboard the USS *Franklin D Roosevelt*, the second of three 45,000-ton Midway class aircraft carriers, which had commissioned in October 1945. Subsequently VF-17A flew further operational trials from the same ship but in early May 1948 they gave an impressive display by carrying out an intensive flying programme simulating operational conditions from the much smaller 14,500-ton USS *Saipan*. Over the course of three days the squadron (now redesignated VF-171) operated at a full strength of sixteen aircraft and flew 176 sorties (i.e. an average of just over ten sorties per aircraft) with no serious incidents, clearly showing that there were no insoluble problems involved in the operation of jet combat aircraft from a carrier deck. Despite this success, the Phantom was never deployed for a full-scale operational cruise and was withdrawn from frontline service with US Navy squadrons in 1949. No other unit actually operated from a carrier, although the aircraft was subsequently used by shore-based Naval Air Reserve units allowing them to gain jet experience. A Marine squadron, VMF-122 commanded by World War II ace Major Marion Carl, also flew Phantoms from November 1947 to

July 1950 and formed an aerobatic display team known as the Flying Leathernecks.

Early jet engines suffered from a number of drawbacks including limited thrust, slow throttle response, and high fuel consumption – particularly at low altitude. Even before the trials with the modified TBF, initial results from early American air force jet prototypes convinced the US Navy that aircraft solely powered by the then available jet engines were not suitable for carrier use. Accordingly it took a cautious approach and issued a specification for a fighter to be powered by a piston engine with a jet as an additional booster for high performance. Nine manufacturers were invited to submit designs and two were eventually awarded contracts to build prototypes, Ryan and Curtiss. The Ryan Model 28 was generally considered the best design and a Letter of Intent dated as early as February 1943 ordered three prototypes under the US Navy designation FR-1 Fireball. In a way this was a surprising outcome as, up to that time, the Ryan Aeronautical Corporation had never built a combat aircraft, being better known for a variety of training aircraft. Nor had they previously built any naval aircraft. Nevertheless, the Fireball was a workmanlike design, being of compact dimensions and incorporating advanced features such as a tricycle undercarriage, a laminar flow wing and flush-riveted skin. The general appearance was remarkably similar to the contemporary Curtiss SC Seahawk floatplane. The single General Electric I-16 turbojet was neatly installed in the rear fuselage with the intakes in the wing roots, the presence of the additional powerplant not being immediately obvious.

In terms of overall performance the Fireball was unexceptional but this was perhaps not surprising as the total installed power was similar to the later versions of the piston-engined F4U Corsair and the composite-powered aircraft was significantly slower. However, the jet engine provided a more sustained rate of climb and the service ceiling of over 43,000 feet was significantly better than the contemporary Bearcats and Corsairs. In addition its compact dimensions made it suitable for use aboard the smaller escort carriers. Even while the three prototypes were under construction the US Navy placed a production contract for 100 FR-1s on 2 December 1943 and orders for a further 600 were placed on 31 January 1945 following successful flight trials. Inevitably these contracts were severely curtailed after VJ-Day and only sixty-six production examples were eventually delivered. It is a tribute to the soundness of the basic design that the production aircraft differed little from the prototypes except that the flap installation was simplified, single slotted replacing double slotted from the fifteenth FR-1 onwards.

To operate the new aircraft a new squadron, VF-66, was formed and this took delivery of its first aircraft in March 1945. Subsequently three aircraft were used for carrier qualification trials aboard the USS *Ranger* on 1 May 1945. Although further tests and trials took place, the squadron was decommissioned on 18 October 1945 and was never permanently based aboard a carrier. However, the aircraft and personnel were transferred to VF-41 (later redesignated VF-1E), which was deployed at sea to take part in exercises aboard various escort carriers. These included USS *Wake Island* in November 1945, USS *Bairoko* in March 1946 and USS *Badoeng Strait* in March and June 1947. Despite the composite power configuration of the Fireball, this activity provided valuable experience in the operation of jet aircraft at sea but its performance was lagging well behind the new pure jets then coming into service and it was withdrawn from use after the last deployment in 1947. In the meantime some of the aircraft were modified for test purposes, including a single XFR-2 fitted with a more powerful R-1820-74W Cyclone piston engine. Other versions included the XFR-3 and XFR-4, which both had the Cyclone engine, and the former also had a 2,000 lb thrust General Electric I-20 turbojet while the latter had an even more powerful 4,200 lb thrust Westinghouse J34 fed via flush intakes set into the sides of the fuselage.

The US Navy had previously ordered a second composite power fighter, this being the Curtiss XF15C-1, which was considerably larger and more powerful than the Fireball. The piston engine was a 2,100 hp Pratt & Whitney R-2800-34W radial engine while the jet was a 2,700 lb thrust Americanised Halford/de Havilland H1-B built by Allis-Chalmers. In general configuration the XF15C was similar to the Ryan FR-1 in that it was a low-wing monoplane with a tricycle undercarriage and the piston engine in the nose but the jet engine installation was different. Like the Ryan it was mounted in the rear fuselage behind the pilot but the tailpipe was very short and the rear lower fuselage was cut away to accommodate this arrangement. Three prototypes were ordered in April 1944 and the first of these flew on 27 February 1945, although initially only as a conventional aircraft without the jet installed. This was later fitted but the aircraft crashed in May and was destroyed before any useful work could be done. The remaining prototypes were completed to the original design, which included a conventional tailplane mounted at the base of the fin, but both were subsequently modified with a substantially larger fin fitted with a long dorsal fillet and the tailplane was mounted on top in a T configuration. While this is a common feature of many modern aircraft, it was very unusual at the time and almost certainly was intended to address serious stability problems in the original layout. In fact, it was not until November 1946 that the handling characteristics and other problems were rectified to the extent that the aircraft could be handed over to the US Navy. Tests showed that the XF15C had a maximum speed of 373 mph at an altitude of 25,300 feet on the piston engine alone and this was boosted to 469 mph with the jet operating. However, by that time it was apparent that the future lay with pure jet aircraft and no production contracts were forthcoming. Sadly this, coupled with the failure of the slightly later Curtiss XP-87 Blackhawk four-engined jet fighter to gain any production orders from the USAF, brought about the end of the once famous Curtiss-Wright as an aircraft manufacturer and the business was sold to North American.

The limited success of the mixed power concept (also tried by the USAAF with aircraft such as the Convair XP-81, which flew in 1944) was not a great setback as the US Navy had already taken steps to supplement the FH-1 Phantoms on order by issuing a specification for a single-engined jet fighter to other manufacturers during 1944 and this led to the award of contracts for prototypes from North American and Chance Vought. The latter's contender was the XF6U-1 Pirate of which three prototypes were ordered in December 1944. The first of these flew in October 1946 and was powered by a single 3,000 lb thrust Westinghouse J34-WE-22 jet engine. The Pirate had a blunt-nosed fuselage housing four 20 mm cannon while the pilot sat high up, well forward of the straight wing. The engine intakes were in the wing roots and jettisonable wingtip tanks were carried. Some thirty production F6U-1s were produced and these were fitted with an afterburning J34-WE-30, giving a maximum thrust of 4,225 lb and a maximum speed of 564 mph at 20,000 feet. Nevertheless, the Pirate offered little advantage over other contemporary jet designs and, in any case, Chance Vought was still fully committed to production of the piston-engined Corsair.

The other aircraft to result from the 1944 specification also had only a limited success and service career but went on to form the basis of one of the most successful jet combat aircraft ever built. This was the North American FJ-1 Fury. As first flown on 27 November 1946, it was a conventional straight-winged jet fighter with a nose intake for the single 3,820 lb thrust Allison J35-A-2 turbojet. North American, of course, was the designer and builder of the P-51D Mustang, probably the best all-round fighter produced in World War II, and the new jet fighter employed many similar components, including the wing and tail assembly. However, the fuselage was completely new with straight-through ducting for the jet via a nose intake and tail exhaust. The J35 had been originally developed by General Electric but had been transferred to Allison for production and further improvement. The need to incorporate fuel tanks in the fuselage led to a rather tubby appearance, although additional fuel was carried in fixed tip tanks. Nevertheless the Fury turned in a respectable performance and one of the proto-types achieved Mach 0.87 during flight testing. The quoted maximum speed of 547 feet at 9,000 feet was well in excess of that of the Phantom, although the rate of climb and service ceiling were substantially less. The armament comprised six 0.5 inch machine-guns but there was no provision for external ordnance. By the time the XFJ-1 prototype flew on 27 November 1946, the US Navy had ordered 100 production FJ-1s but this order was subsequently cut to only thirty-nine aircraft as it became apparent that other types offered even greater perform-ance. However, within its performance limitations the Fury proved to be a docile aircraft to fly and well suited to carrier operations. Consequently an operational squadron, VF-5A under the command of Cdr Evan P Aurand USN, was formed in 1948 and began carrier qualification aboard the USS *Boxer* in March 1948. Subsequently VF-5A formed part of the ship's air wing until the end of 1949 when the remaining Furies were transferred to Reserve units. This was the first time that a jet fighter squadron had been permanently assigned to a carrier and marked a notable step forward for naval aviation.

In the meantime the jet-powered Phantom had entered service but its operational career, as already related, was very short and spanned only a couple of years. The reason for this was that its performance, while in advance of naval piston-engined fighters, lagged behind that of the land-based jet fighters then coming into service with the USAAF and with potential adversaries. In traditional style, the US Navy therefore ordered a Phantom replacement even before that aircraft had entered squadron service. Amongst other contenders, McDonnell was given the chance in 1945 to design a successor and a contract for three XF2D-1s was awarded. The designation was later changed to F2H-1 and the name Banshee was selected (perhaps appropriate given the intense high-pitched noise of the early jet engines). The new design was basically an enlarged Phantom fitted with two 3,000 lb thrust Westinghouse J34 axial flow turbojets. This was almost twice the installed power available to the Phantom and much of the lengthened fuselage was taken up with additional fuel tankage, capacity rising from 375 US gallons in the FH-1 to 526 US gallons in the F2H-1. A much more effective armament of four 20 mm cannon was fitted below the nose (as opposed to the four machine-guns which, in the Phantom, were mounted in the upper nose where flash was a serious problem at night). The prototype XF2H-1 flew at McDonnell's St Louis facility on 11 January 1947 flown by test pilot Robert M Eldholm. Flight tests and initial carrier trials revealed no major problems so that production F2H-1s differed only in having a slightly longer fuselage (increased from 39 feet to 40 feet 2 inches) and a tailplane without dihedral. Fifty-six aircraft were delivered between August 1948 and August 1949. However, the major production version was the F2H-2, which was fitted with wingtip tanks of 200 US gallons capacity and more powerful 3,250 lb thrust J34-WE-34 turbojets, which allowed the normal loaded weight to rise from 16,200 lb to 20,555 lb. Apart from the additional fuel, this increase allowed the carriage of two 500 lb bombs or six 5 inch HVAR rockets. Some 306 standard F2H-2s were built but this version formed the basis of a number of specialist variants that were also produced in small numbers. Of these, perhaps the most interesting was the F2H-2B, which was specially developed for the nuclear strike role, something in which the US Navy invested a considerable of effort and money (see Chapter 8). In the case of the Banshee the wing was strengthened to allow the carriage of a single Mk.7 1,650 lb or Mk.8 3,230 lb nuclear bomb under the port wing and twenty-seven of this version were built. Other variants included the single-seat F2H-2N night fighter with an AN/APS-19 radar in an extended nose (three conversions and eleven new build), and the F2H-2P, which had all armament removed and carried no fewer than six vertical and oblique cameras in an extended and widened nose. Eighty-nine F2H-2Ps were built, the last being delivered on 28 May 1952.

Only two months previously, on 29 March 1952, the first F2H-3 had flown. This represented a major step forward as it was the US Navy's first purpose-designed all-weather fighter and built on experience gained with the F2H-2N. The new version had a substantially lengthened fuselage (now measuring 48 feet 2 inches), which gave space for a Westinghouse AN/APQ-41 radar in the nose and allowed the total internal fuel capacity to rise to 1,102 US gallons. The now optional tip tanks were slightly reduced in size, carrying only 170 US gallons each. Despite the nose radar, space was found to increase the ammunition supply (220 or 250 rounds per gun). Hardpoints were fitted for the carriage of four 500 lb bombs or several HVAR rockets and eventually the Sidewinder air-to-air guided missile could be carried, although these did not become available until 1956. Apart from the lengthened fuselage, the F2H-3 could be distinguished by a dihedral tailplane and in service the leading edges of the tailplane were also extended. A total of 250 were built. The final production version was the F2H-4, of which the last of 150 was delivered on 24 September 1953. This was basically the same as the -3 but was fitted with a

Hughes AN/APG-37 radar and was powered by 3,600 lb thrust J34-WE-38 engines.

The Banshee, or Banjo as it was nicknamed by US Navy pilots, served for over a decade and the fighter's potential was given an early demonstration when an F2H-1 reached an altitude of 52,000 feet (an unofficial world record, mainly carried out in an attempt to discredit the Air Force's B-36 bombers, which had hitherto claimed immunity from fighter inter-ception at such altitudes). A less flattering achievement was Lt JL Fruion's escape from an F2H-1, which went out of control at 30,000 feet, in the process becoming the first US pilot to successfully use an ejection seat in an emergency situation. As already mentioned, VF-171 was the first unit to fly Banshees but the aircraft flew with several other US Navy and Marine squadrons and even when supplanted as a day fighter from 1953 onwards, it remained in its -3 and -4 versions as the Fleet's standard all-weather fighter until around 1959. As an aside, it is interesting to note that thirty-nine F2H-3s were transferred to the Canadian Navy in between 1955 and 1958 and remained in service until 1962, by which time a significant number had been lost in accidents due to the problems of operating these relatively advanced jets from the decks of the ex-British Magnificent class light fleet carriers.

McDonnell's early experience in the post-war years with the original Phantom and its successor the Banshee gave the com-pany a commanding lead as a supplier of jet fighters to the US Navy, but they were to be closely pressed by the traditional supplier of naval fighters – Grumman. This company had lost out in the early development of jet fighters as they were heavily committed to the production and development of piston-engined fighters such as the Hellcat, Bearcat and Tigercat. Even when they obtained a contract for a jet fighter, this turned out to be something of a blind alley as the selected engine was the low-powered 1,500 lb thrust Westinghouse J30 and the Grumman team calculated that four of these, buried in the wings, would be required to achieve the required performance. Like McDonnell with the Phantom, they soon found that such an arrangement was complex and inefficient and the search began for an alternative. This was where the previous association with

British engines bore fruit and when the contract for the two-seat XF9F-1 was cancelled, Grumman had already commenced design work on a single-seat fighter to be powered by a variety of US and British engines, including a twin Derwent project. However, by this time details of the 5,000 lb thrust centrifugal flow Rolls-Royce Nene became available and arrangements were being made for this to be produced in America by the Taylor Turbine Corporation, although eventually it was built as the J42 by Pratt & Whitney.

The prototype XF9F-2 Panthers were actually powered by imported Rolls-Royce Nenes pending availability of the US-built versions and the first flight took place on 21 November 1947. The Panther was a straight-winged design with a streamlined fuselage, although in order to keep jet pipe length to a minimum, the rear fuselage was cut away under the tail assembly. The engine was fed by wing root air intakes. Although not fitted to the prototype, all subsequent Panthers had fixed 120 US gallon wingtip tanks. The initial armament was four 20 mm cannon with 190 rounds per gun and it was not until the aircraft had entered service that modifications were incorporated to allow the carriage of up to 3,000 lb of bombs or other ordnance (such modified aircraft were initially designated F9F-2B). Initial flight trials revealed a number of problems, including a lack of directional stability (an important factor as this affected the aircraft's suitability as a gun platform) and longitudinal instability at low speeds. The latter problem was caused by fuel slopping around the fuel tanks and was solved by fitting internal baffles, while an increase in fin and rudder area partially cured the directional problems. Compared with other contemporary naval aircraft, the Panther's stalling speed was on the high side and it was certainly something of a 'hot ship' for its day. In compensation its performance was significantly better than that of the Phantom or the Fury with a top speed of 575 mph at sea level, a rate of climb in excess of 5,000 feet/min and a service ceiling of 44,600 feet.

Initially the Phantom was ordered in two versions, the F9F-2 with the J42 (Nene) and the F9F-3 with the 4,600 lb thrust Allison J33. Comparative tests soon showed the superiority of the Nene-engined version and most of the fifty-four F9F-3s delivered were re-engined with the J42. Total production of the F9F-2 reached 594 aircraft, which were delivered between May 1949 and the end of 1950. It was replaced on the production lines by the F9F-4 and -5. These two variants were again to be powered by different engines, the 6,250 lb thrust Allison J33-A-16 turbojet in the case of the -4 and the similarly rated Pratt & Whitney J48-P-6 for the -5. The latter engine was an Americanised version of the Rolls-Royce Tay and once again proved superior in service. Consequently, most of the 109 F9F-4s delivered were either re-engined in service or completed at the factory with the J48. The definitive F9F-5 was the most important production version of the Panther with no fewer than 616 being delivered as well as a further thirty-six F9F-5P photographic reconnaissance variants. (Previously a few F9F-2s had been converted to this role as the F9F-2P in 1950 at the outbreak of the Korean War.) Apart from the more powerful engines, the F9F-4 and -5 shared a fuselage lengthened by 19.5 inches ahead of the wing in order allow additional fuel to be carried and this required an increase in the size and area of the tail fin to retain directional stability. The wing hardpoints were strengthened to allow a total bomb load of 3,450 lb to be carried. Water injection boosted the thrust rating of the J48 to 7,000 lb and allowed the F9F-5 to become the US Navy's first fighter capable of exceeding 600 mph at sea level. However, the maximum weight had risen from 9,300 lb to 10,150 lb and consequently the rate of climb remained substantially unaltered while the service ceiling was reduced to 42,800 feet.

Entering service in 1949, the Panther equipped two US Navy and one Marine squadron, as well as the Blue Angels demonstration team, by the end of the year. When the Korean War broke out in mid 1950 the US Navy had eight frontline jet fighter squadrons in service, of which six flew Panthers and the remaining two were equipped with Banshees. Two squadrons (VF-51 and VF-52) were aboard the USS *Valley Forge*, which became the first US carrier to fly operational missions. The Panther thus became the first US Navy jet to fly in combat, an early success being the shooting down of a Yak-9 piston-engined fighter on 3 July 1950. However, on 1 November 1950 the first Chinese MiG-15s were encountered and the superior performance of the Russian-built jet was immediately apparent, forcing the US Navy Panther pilots to rely on their superior training and experience to even things up. In the first US Navy jet versus jet combat on 9 November 1950, Cdr WT Amen commanding VF-111 aboard the USS *Philippine Sea* managed to destroy a MiG-15 and in the course of the Korean War Panthers shot down another four with no loss to themselves.

It is interesting to examine the reasons for the MiG-15's superiority, particularly as it was effectively powered by the same engine! This derived from the fact that the Socialist British government had, in a demonstration of solidarity with its erstwhile wartime allies, gifted examples of the Rolls-Royce Derwent and Nene jet engines together with all technical details in early 1947. This naive gift gave the Soviet aero engine industry a tremendous boost and the Nene was immediately put into production as the VK-1, later becoming the RD-45, and one was available to power the prototype MiG-15, which first flew in July 1947. However, the MiG also benefited from analysis of captured German material at the end of World War II when Soviet designers and engineers quickly became aware of the benefits of sweptback wings. The MiG-15 was designed from the start with a highly swept wing (35 degrees at quarter chord) and this was the key to its success. American engineers had also gained access to the same data but for once they were relatively slow off the mark and the first US swept-wing combat aircraft (the F-86 Sabre, derived from the US Navy's FJ-1 Fury) did not fly until almost a year after the MiG prototype. The appearance of the MiG-15 in Korea laid new urgency on the development of swept-wing fighters for both the USAF and the US Navy and it was clear that any new aircraft would have to adopt this configuration.

As already mentioned, by mid 1950 the US Navy had eight jet fighter squadrons in service aboard carriers and the majority of these were powered by British-designed jet engines. Given that the Royal Navy had pioneered the operation of jets from carrier decks and that it was supported by the world's best aero engine industry, it could have been expected that at least a similar stage of development might have been reached. Sadly the actual state of affairs was almost as bad as it could be. In fact, the Royal Navy did not deploy an operational jet fighter squadron at sea until 1952 and none were available for deployment to Korea. As will be seen, there were many reasons for this but the main one can be traced back to the history of the Fleet Air Arm, which had only come fully under naval control in 1937, having previously operated as a branch of the RAF. Even a decade later the ramifications of this arrangement were still being felt as there were still many senior officers who lacked carrier or aviation experience and the drawing up of specifications and liaison with manufacturers was done through the offices of a central Ministry of Supply. Compare this with the US Navy whose own air arm had been an integral part of the Navy from the day it was formed and who had their own Bureau of Aeronautics (running in parallel with the Bureau of Ships) to oversee the design and production of naval aircraft. Thus when the US Navy entered the jet age they were able to issue specifications for purely naval aircraft, which were consequently designed from the outset to operate from aircraft carriers. In Britain the Admiralty was much less pro-active and generally issued specifications for naval aircraft that were based on already existing projects for land-based aircraft initially designed for the RAF. In addition the Admirality tended to be overly cautious and looked for incremental improvements in performance rather than risking commissioning entirely new and more advanced designs.

In the event the Royal Navy turned to its traditional fighter suppliers, Hawker and Supermarine, for its first jet designs but in both cases the evolutionary approach led to very long development times and slow entry into service. While this was partly due to the official attitudes of the time, it must also be recognised that funding for such ambitious projects was in short supply as Britain struggled to get its shattered economy back into shape after six long years of war. The Royal Navy's first fully operational jet fighter, as opposed to the Sea Vampire, which was always regarded as an interim type and never equipped a frontline squadron, was the Supermarine Attacker. This clean-looking, straight-winged aircraft could trace its ancestry directly back to the piston-engined Spitfires and Seafires, which for so long had provided the backbone of the

British fighter force. As has already been related, post-war developments led to the Spiteful and Seafang whose main advance was a new laminar flow high-performance wing. As early as the summer of 1944 Supermarine was asked to begin design work on a new fighter designed around one of the new jet engines, in this case the RB41, which eventually developed into the highly successful 4,500 lb thrust Nene. The new aircraft was designated Type 477 and utilised the Spiteful-type wings, including the armament of four 20 mm cannon, married to a new fuselage in which the Nene engine was fed by lateral air intakes while the pilot sat well forward in the nose. Unusually for a jet aircraft, a conventional tailwheel undercarriage was provided. Specification E.10/44 was written around the Supermarine proposals and three prototypes were ordered on 5 August 1944. Evidence of early and positive naval interest in the new Type 477 is illustrated by the fact that two of the prototypes were semi navalised with long stroke undercarriages and arrester hooks, although folding wings were not provided at this stage. Subsequently a further eighteen naval examples were ordered under specification E.1/45 but the flight of the prototype (TS409) was delayed for several months while problems with the new wing were investigated in the Spiteful/Seafang test programme. One of the major issues was the behaviour of the ailerons at high speeds but other control-related problems dogged the programme, which continued after the first flight of TS409 on 27 July 1946. The delays resulting from the exhaustive flight test programme resulted in the Admiralty cancelling its order for the Type 477 and instead procuring eighteen Sea Vampire F.Mk.20s.

Following the initial flight trials, the second prototype and first navalised Type 477 flew in June 1947 and featured several improvements, including redesigned tail surfaces, modified flaps and spoilers, additional fuel capacity and, importantly, a Martin-Baker ejector seat. After further modifications the aircraft, now officially known as the Attacker, was put through a series of airfield dummy deck landings (ADDLs) before beginning deck landing trials aboard HMS *Illustrious* at the end of October 1948. These demonstrated that the tailwheel under-carriage configuration had certain advantages for a naval aircraft at that time as the nose-up landing attitude gave rise to an aerodynamic braking effect, which helped to slow down the aircraft on final approach. It was also more suited to a catapult launch in which the tail-down attitude provided greater lift as the aircraft accelerated. As a result of the shipboard trials further modifications were made to the third prototype and eventually an order for sixty Attacker F.1s was forthcoming from the Admiralty in November 1949 and the first production aircraft flew on 5 April 1950. By the end of August enough air- craft had been delivered to equip 800 Squadron, which thus became the Royal Navy's first frontline jet fighter squadron, although they did not go to sea until March 1952 when the new carrier HMS *Eagle* commissioned.

The initial sixty Attacker F.1s were later fitted with an extended dorsal fin to improve directional stability and this became standard on all subsequent aircraft, which were also fitted to carry a 250-gallon underfuselage drop tank. Provision to carry two 1,000 lb bombs resulted in the FB.1. Eight of these were built before production turned to the FB.2 variant, which had a more powerful Nene 102 engine, an improved cockpit canopy and other modifications. This was the major variant and a total of eighty-four were built, the last being delivered in 1953. Two other frontline squadrons (803 and 890) received Attackers but neither was deployed aboard a carrier other than for brief periods of training and the aircraft's period of service was relatively short. All three operational squadrons had converted to other types by the end of 1954, although the Attacker was used by training and reserve squadrons until 1957.

Although the aircraft resulted initially from an RAF specifi- cation for a day fighter, that service rapidly lost interest and placed no orders so that all subsequent development was concentrated on the naval version (however, some thirty-six Attackers were sold to the Pakistan Air Force). Due to its pro- tracted development time, the Attacker was already obsolescent by the time it entered service, although on paper its perform- ance compared favourably with the contemporary Grumman F9F Panther. The difference was that the latter had entered service two years earlier and was available in significant numbers for operational service when the Korean War broke out in 1950.

The Royal Navy's main jet fighter in the 1950s was the Hawker Seahawk but like many British aircraft of the period it had suffered from a long gestation period. In fact, its origins went back to 1944 when the Attacker was also first conceived. In the case of the Hawker aircraft the company made an initial proposal for a single-seat jet fighter based on the piston-engined Fury but powered by a RB.41 Nene jet engine. As work progressed Hawker evolved the idea of a bifurcated split exhaust, which vented either side of the fuselage, instead of a conventional straight-through jet pipe. At the time it was thought that this configuration would reduce thrust losses in the jet pipe and it also had the incidental benefit of providing additional space for fuel tankage in the rear fuselage. By October 1945 the design was sufficiently refined for Hawker to begin work on construction of a prototype under the company designation P.1040. By this time RAF interest had evaporated as the aircraft offered little advance in performance compared with the twin-engined Gloster Meteor but the Admiralty was keen to develop it as a fleet fighter and issued Specification N.7/46, authorising the construction of three prototypes. The first of these flew on 2 September 1947 but was only an aerodynamic prototype, which nevertheless contributed immensely to the development programme and resulted in several refinements being incorporated in the fully navalised second prototype, which flew on 3 September 1947. This featured folding wings, was fitted for catapult launching and carried the full armament of four 20 mm cannon in the lower nose. After initial trials at Boscombe Down in April 1949, this aircraft then carried out deck landing trials aboard HMS *Illustrious*. The third prototype flew in October 1949 and subsequently orders were placed for an initial 150 Seahawk F.1s, the first of these being delivered in November 1951. In fact, Hawker only built thirty-five Mk.1s, all subsequent production and further development being transferred to Armstrong Whitworth at Coventry. Some ninety-five Mk.1s were produced, these being superseded by forty Seahawk F.Mk.2s. This model introduced powered flying controls, which were found necessary to overcome aileron problems at high speed (echoing experience with the Spiteful and Attacker). While both early versions could carry 90-gallon drop tanks, the later FB.3 was modified to permit the carriage of two 500 lb bombs and first flew in March 1954, a total of 116 being delivered. Later variants included the F.G.A.4, which could carry up to four 500 lb bombs and the F.B.5 and F.G.A6, which were basically Mk.3/4 Seahawks with an uprated 5,200 lb thrust Nene 103 engine.

Unlike the Attacker, the Seahawk found a ready export market and customers included the Netherlands Navy (twenty-two aircraft), West Germany (sixty-four) and India (twenty-four). In addition, the Indian Navy received a further twelve ex-Royal Navy aircraft and twenty-eight ex-German Seahawks. In all, some 554 Seahawks were produced in addition to the three prototypes. Many of these orders resulted from the aircraft being accorded 'super priority' status during the Korean War (along with the Hunter and Canberra for the RAF). Nevertheless, the first Royal Navy Seahawk squadron (806 NAS) did not form until March 1953 and did not embark on HMS *Eagle* until February 1954, well after the end of hostilities in the Far East. Subsequently, the Seahawk formed the backbone of the Royal Navy fighter strength until the end of the 1950s, equipping no fewer than thirteen frontline squadrons, three reserve squadrons and five training units. Although it was too late for Korean service it did see action in the Anglo-French Suez operations in the autumn of 1956 when no fewer than seven Seahawk squadrons were involved, flying from three carriers (*Eagle*, *Albion* and *Bulwark*). Well liked by its pilots, the Seahawk was perhaps one of the most graceful jet combat aircraft ever built and it was certainly a useful ground attack aircraft. However, there is no denying that by the time it entered service it was a whole generation behind what was available to the US Navy.

At the time of the Suez operations the Seahawk's companion in the carrier air groups was the de Havilland Sea Venom, a two-seat, radar-equipped, single-engined all-weather fighter. The need for radar-equipped night fighters had been identified during World War II and had been met initially by fitting radar pods to single-engined aircraft such as the F6F-5N Hellcat or the Firefly NF.1. Subsequently, as already described, a two-seat version of the twin-engined Sea Hornet saw service as the

NF.Mk.21 but the need for a jet-powered all-weather inter-ceptor led to the adoption of a variant of the de Havilland Venom. The original single-seat Venom FB.Mk.1 Venom had been developed from the earlier Vampire, the main external difference being the introduction of a new thin profile wing with a swept leading edge. Wingtip fuel tanks were also a standard fitting on all Venom variants. Power was provided by a de Havilland Ghost 103 turbojet initially rated at 4,850 lb thrust, but this was later increased to 5,200 lb in the Ghost 105. De Havilland had already produced a two-seat version of the Vampire, both as a night fighter and as an advanced trainer, and so it was natural that the company should do the same with the Venom. The prototype Venom NF.Mk.2 was actually a private venture and first flew in August 1950, but it was subsequently ordered in quantity for the RAF. The Royal Navy also evaluated the NF.Mk.2 prototype and as a result ordered a navalised version under Specification N.107, which was designated Sea Venom NF.Mk.20 and first flew early in 1951. Carrier trials took place aboard HMS *Illustrious* in July 1951 and although the first two examples did not have folding wings, this feature with power operation was incorporated in the third prototype and all subsequent production aircraft. The first of fifty production aircraft (now designated FAW.Mk.20) flew on 27 March 1953. Subsequent versions included the FAW.Mk.21 (167 built) with power ailerons, improved canopy, long stroke undercarriage, Martin-Baker ejector seats for both crew members and a new radar, and the FAW.Mk.22 with an uprated Ghost 105 engine and A.I.Mk.22 radar (thirty-nine built).

The Sea Venom entered service with 890 Squadron in March 1954 and subsequently equipped nine frontline and four second line squadrons. Although outside the scope of this book, the Sea Venom FAW.Mk.21 became the Royal Navy's first missile-equipped intercepter in 1958 when aircraft of 893 Squadron carried out firing trials with the Firestreak missile while embarked in HMS *Victorious*. The Sea Venom was also selected by the Royal Australian Navy and thirty-nine were eventually delivered from 1955 onwards. A more interesting export success was the French Navy being supplied with four Sea Venom Mk.20s assembled in France by SNCA de Sud-Est, these being followed by a single French-built prototype. Subsequently Sud-Est produced seventy-five aircraft as the Aquilon 202 for service with the French Navy. Overall, the Sea Venom fulfilled its required role but suffered from having only a single engine (contemporary RAF all-weather fighters were all twin-engined) and the cockpit was extremely cramped with limited space for additional equipment.

The US Navy also had a requirement for a two-crew, jet-powered night fighter and this was to be met by the Douglas F3D Skyknight. This project had its origins in 1945 when the Bureau of Aeronautics raised a requirement for a two-seat carrier-based radar-equipped jet fighter capable of intercepting hostile aircraft, flying at a speed of 500 mph and an altitude of 40,000 feet at a range of 125 miles. Proposals from various manufacturers resulted in contracts being awarded to Douglas for the XF3D-1 and Grumman for the XF9F-1. In the event the Grumman design evolved into a single-seat aircraft resulting in the Panther so Douglas had the field to itself. The F3D Skyknight was a relatively conservative design with a mid-mounted straight wing and the two crew members seated side-by-side in the nose cockpit. In order to carry the two crew, an armament of four 20 mm cannon, radar equipment and sufficient fuel, a twin-engined configuration was adopted in which the 3,000 lb thrust Westinghouse J34-WE-24 turbojets were mounted semi-externally under the fuselage centre section. The result was an extremely large aircraft by the standards of the time and with a maximum all-up weight of 21,500 lb it was one of the heaviest. The first flight was on 23 March 1948 and no serious problems were encountered in the test flight programme so an initial order for twenty-eight F3D-1s was placed in June of that year. Deliveries began in February 1951 with aircraft initially going to Composite Squadron VC-3. A total of 268 Skyknights were built, the majority being the F3D-2, which was to have been powered by the larger and more powerful Westinghouse J46 but problems with this engine resulted in reversion to the J34-WE-36 rated at 3,400 lb thrust.

In service, the Skyknight was operated almost exclusively by US Marine Corps squadrons and saw very little carrier-based service. Despite that, it was remarkably successful in the Korean

War, destroying more enemy aircraft than any other US Navy or Marine aircraft and scoring the first jet versus jet night combat success when a MiG-15 was shot down on 2 November 1952 by a Skyknight from VMF(N)-513 based at Kunsan. The Skyknight also had a relatively long career, examples equipped for electronic warfare seeing service in the Vietnam War until retired in 1969, although these activities are outside the scope of this book.

The aircraft described in this chapter almost without exception had their origins in the closing stages of World War II, or in the immediate post-war era. All employed conventional straight wings and were designed around early jet engines of limited thrust. From 1946 onwards, the results of German wartime research and experience became available and pointed to the benefits of swept wings and other more exotic configurations as a means of overcoming the problems associated with high-speed flight in the transonic region. Continuing engine development also produced more powerful and efficient engines with thrust ratings rising from around 3,000/4,000 lb to 6,000/7,000 lb with more to come. As these advances filtered through to the aircraft designers a whole range of new naval combat aircraft was produced and began to enter service in the early 1950s. However, these were larger, heavier and faster than their predecessors and used more fuel and could carry a greater load of more complex weapons. Taken together these characteristics placed new demands on the aircraft carriers from which they were to operate and solutions to the many associated problems became a matter of some urgency.

5

ADAPTING THE CARRIERS – 1950 TO 1955

Although by 1950 the US Navy was clearly the world's most powerful navy with a carrier fleet to match, the Royal Navy was making up for its smaller size by pioneering a series of vital inventions that were to prove the key to operating the new generation of high-performance jets then under development. However, being a pioneer in any field is never easy and many ideas were tried and tested before a concept could be brought to the stage where it could be safely applied to everyday carrier operations. Inevitably, some of these ideas led to a blind alley, but not before a considerable amount of time and effort had been expended, and in this category is firmly placed the concept of the flexible deck.

This had its origins as far back as 1944/5 when the first jet fighters were entering service. It was immediately apparent that they lacked the range of their piston engine contemporaries due to the high fuel consumption of jet engines. One idea put forward was that if the undercarriage could be dispensed with then the saving in weight and space could be utilised to substantially increase fuel tankage. The idea was not quite as bizarre as it might appear as several advanced German aircraft such as the Me.163 rocket-powered interceptor and early versions of the Arado Ar.234 jet bomber had used this method of operation. The main drawback was obviously the necessity to lift the aircraft onto some form of cradle after landing before it could be moved. For this reason the RAF quickly abandoned the idea but the Royal Navy was more interested as it saw it as one way of redressing the weight penalty inherent in naval aircraft necessitated by the need for extra equipment and strengthening for deck operations. Also, as naval aircraft were routinely launched from catapults, this method could easily be applied to undercarriage-less aircraft mounted on a suitable trolley. Major Green, one of the engineers working at RAE Farnborough, put forward a scheme whereby a landing aircraft could engage an arrester wire and be brought to rest on a large rubber mat suspended above a carrier's flight deck. After a successful landing, the aircraft would be winched to the forward end of the mat where it would be pulled onto a specially designed trolley. It could then easily be moved around the deck and the same trolley could be attached to a catapult for take-off. It was estimated that a landing rate of one aircraft every thirty seconds could be achieved.

Further development now became the responsibility of the Naval Aircraft Department at Farnborough who refined the design to include a series of inflated canvas tubes to provide an air bed under the rubber mat in order to absorb and cushion the impact of the landing aircraft. By 1946 an experimental bed had been set up at Farnborough and a series of experiments was carried out in which surplus Hotspur glider fuselages loaded with various amounts of ballast were dropped onto the bed. It was found that the best results were obtained with a combination of the inflated tubes laid out under a tensioned rubber mat. (In the typical make-do fashion of many British projects, the inflatable tubes were actually made from surplus fire hoses!) Further tests were then made with unmanned complete Hotspur airframes, which were rocket launched so that they engaged the arrester wire almost as soon as they achieved flight. In order for a manned aircraft to make a successful engagement, it would be necessary for the pilot to approach the deck and its single arrester wire (suspended some three feet above the deck) at a fairly flat angle and at a speed that would not impart too much energy to the system on landing. Also, it would be necessary to judge height over the deck very accurately – typically a height of five feet was required. A series of tests was therefore flown by the famous naval test pilot Cdr Eric Brown RN who had already made the first successful carrier jet landing in December 1945, using an Aircobra and then a Hellcat. These tests consisted of flying typical approach profiles and then coming over the deck, although no attempt was made to engage the arrester wire, which was not rigged. It was quickly apparent that such flying techniques presented no particular problems as was confirmed when similar trials were flown by a representative selection of other pilots.

All was now ready for the first actual landing on the rubber deck, which was installed on the airfield at Farnborough alongside one of the runways. The mat was 200 feet long and 60 feet wide, with steel ramps at either end and an arrester wire suspended 30 inches above the landing end of the deck. The aircraft selected was the second prototype Sea Vampire F.20 and for the trial it retained its normal undercarriage, although, of course, this was left retracted. On 29 December 1947 Cdr Brown made the first attempt at a rubber deck landing in a piloted jet aircraft. Unfortunately this attempt was not a resounding success as wind eddies on the final approach caused the aircraft to sink too low and, as the pilot attempted to pull up, the tail booms and arrester hook hit the lead-in steel ramp causing the aircraft to pitch forward and the arrester hook to lock in the UP position. Consequently the wire was not engaged and the aircraft pitched nose down into the deck, bounced up, hit the steel ramp at the upwind end and finally skidded to a halt on the grass area beyond, having incurred substantial damage to the nose and cockpit area. Fortunately Cdr Brown was unhurt and after some modifications to a second Sea Vampire and to the method of approaching the deck, a successful landing was made on 17 March 1948. By November of that year a total of thirty-three successful landings had been completed and only a few minor, easily rectified, problems had been experienced.

In the light of the promise of the early trials, authorisation was given for seagoing trials to be carried out aboard the light fleet carrier HMS *Warrior*. The carrier was fitted with a 300-foot rubber deck during the middle part of 1948 and this work was completed in time for trials to commence at the beginning of November. The installation was placed on the after end of *Warrior*'s flight deck, leaving the forward 300 feet free for take-off and parking. The first successful landing was made on 3 November and subsequently over 150 deck landings were made, almost totally without incident and by a wide cross-section of both Royal Navy and RAF pilots as well as one US Navy pilot. The aircraft used in all these landings except the first were standard Sea Vampire F.21s, which were capable of being catapult launched. (The first landing used an F.20, which could not be catapulted and required a 42-knot wind over the deck before it could be flown off from the forward portion of the ship's flight deck.) In all, 271 landings were made using both shore-based and shipboard installations and the fact that there were no accidents clearly illustrated that the system worked. Unfortunately, aircraft developments had to some extent over-shadowed the original idea behind the concept and its wide-spread adoption would require a substantial investment in deck facilities both aboard ship and at shore bases, and the necessary finance was just not available.

As a postscript, the idea was revived in early 1950 and a 400-foot deck was erected at Farnborough. Trials again utilised Sea Vampires but the prototype Sea Hawk made two successful landings in 1953 and in all some 304 landings were made between 1952 and 1955. The impetus for these new trials came from the US Marine Corps, which saw the system, together with zero-length rocket launchers for take-off, as being a way of getting high-performance jets ashore in the immediate after-math of an amphibious assault. To further the project a flexible deck was erected at the US Navy test establishment at Patuxent River where trials were conducted with a number of US aircraft including the F9F Panther. Ultimately, the Marine requirement was to be met more than a decade later by a new generation of VTOL (Vertical Take Off and Landing) aircraft in the shape of the Harrier, but this is outside the scope of this book.

Although work on the rubber deck finally petered out, the system did have one advantage over conventional deck operations, which partly accounted for the fact that the Royal Navy persevered with the trials. In the 1940s any aircraft that failed to engage the arrester wires would either end up hitting the crash barrier amidships, or missing both wires and barrier and ending up crashing into the forward deck park with calamitous results. On the other hand, because of the technique employed in landing on the rubber decks, in which the aircraft was flown straight and level at a constant airspeed on the final stages of the approach, if it missed the wire it was a simple matter for the pilot to continue to overfly the deck and other aircraft before climbing away for another approach. Such a facility was obviously desirable and a more elegant and simple solution was eventually discovered, partly stemming from the rubber deck concept. In August 1951 a conference was set up to consider how undercarriage-less aircraft could be operated at sea and the rubber deck featured prominently in the discussions, which were chaired by Captain Campbell RN. It was agreed that the elimination of the crash barrier was a very useful attribute but in order to provide more protection for the forward deck park, Campbell made a proposal that the rubber deck be angled some 10 degrees off the ship's centreline and he produced a hand-drawn sketch to illustrate his ideas. At the time it was accepted as one of several possible solutions but no decision was taken at the conference. However, one of the delegates was an RAE scientist, Lewis Boddington, who had been involved in the trials at Farnborough and aboard HMS *Warrior*. After giving the matter some thought he realised that the principle of angling the landing deck away from the axis of the ship could actually be applied to ordinary carrier landings by conventional aircraft. Within three weeks he had drawn up plans to adapt the new carrier *Ark Royal* to accommodate a 10-degree angled deck, which he presented to the Director of Naval Construction. While this was being considered, Boddington and Campbell arranged for the light fleet carrier HMS *Triumph* to undertake trials of the new concept. Modification of the ship was limited to removing the flight deck's axial white centreline markings and replacing them with a new centreline offset 10 degrees to port. The arrester wires were left at their normal alignment as they were not required for the intended demonstrations, which consisted of touch and go landings only. The trials were a great success and the concept obviously had the potential to completely revolutionise flight deck operations, as well as making a major contribution to flight safety.

Despite this, the ever-cautious Admiralty while recognising the implications of the system, initially decided that it was not worth modifying existing carriers but would wait until a new generation of carriers was ordered. However, this view was soon modified by the actions of the US Navy who, as will be related, recognised the angled deck as a heaven-sent solution to a whole series of problems associated with the operation of heavy jets at sea. By the beginning of 1953 they had completed conversion of the Essex class carrier USS *Antietam* (CV36) to incorporate a fully angled deck and the ship was generously made available to the Royal Navy in the summer of that year for trials in the English Channel. The results fully confirmed the benefits of the angled deck and the Admiralty was convinced of the necessity to adapt existing ships and to so modify those still under construction. The first Royal Navy carrier to be equipped with an angled deck was HMS *Centaur*, the first of a class of three ships developed from the Colossus and Majestic class light fleet carriers. *Centaur* had been completed in September 1953 but almost immediately was taken in hand for the necessary modifications, which, in her case, were limited to painting a deck centreline offset by 5.5 degrees and realigning the arrester wires. A modest deck edge extension was added amidships on the port side, necessitating the removal of three twin 40 mm mountings. In this form the ship was back in service in the autumn of 1954, although her initial air group comprised piston-engined Sea Furies and Avengers, the first jet squadrons not embarking until 1955 (803 and 806 Squadrons equipped with Seahawks). The first British carrier to have an angled deck incorporated while under construction was HMS *Ark Royal* whose completion was delayed by two years to allow this and other new equipment to be fitted. Launched in May 1950, she consequently did not commission until February 1955.

Amongst the new equipment installed in *Ark Royal* was another British invention – the steam catapult. At the end of World War II all British carriers of the Illustrious and Colossus/Majestic classes were fitted with BH.III hydro-pneumatic catapults. In these installations the power was provided by compressed air driving a piston within a cylinder. The piston's movement was transferred to a launch trolley running in a track set in the flight deck by means of a complex system of wires and pulleys, giving an 8:1 increase in speed and travel. The trolley track in all BH.III installations was 140 feet 9 inches long and on completion of the launch the whole moving mass was brought to a halt hydraulically, the piston entering a water-filled cylinder and forcing the liquid out through apertures. Hydraulic power was then used to push the piston back to its start position ready for the next launch. Incredibly, the moving parts of the catapult weighed no less than 15,000 lb, as much as most of the aircraft that were being launched. The final versions of the BH.III could accelerate a 20,000 lb aircraft to a speed of 66 knots but maintenance was a major issue as the wires needed replacing after approximately 900 launches and their replacement could take up to two days' work. Further development of this type of catapult resulted in the BH.V, which could accelerate a 30,000 lb aircraft to 85 knots and these were intended for the Majestic, Centaur and Ark Royal class carriers.

It is not often realised that although these catapults were hydro-pneumatic in operation, the power to compress air and drive the hydraulic pumps was derived from steam turbines, which accounted for a significant proportion of the available boiler output when a rapid series of launches was made. As early as 1936 Mr C Mitchell, an engineer working with MacTaggart Scott who produced the Royal Navy's catapults, proposed a much simpler form of catapult in which gas pressure (steam or air) drives a piston along a tube. Attached to the piston was a traveller that projected through a slot in the top of the piston, gas losses being prevented by fitting a flexible seal, which was opened and closed as the piston passed. At the time the idea was thought impracticable (although Brunel had used it in his atmospheric railway in the 1840s) and no further work was done until 1944 when Mitchell (then a Commander (E) RNVR) inspected German V-1 launch sites and realised that the catapult was very similar to his own patented design.

Trials with a captured German catapult re-assembled at Shoeburyness showed its potential to launch the heaviest naval aircraft and development of a shipboard steam catapult was authorised in 1946. After various trials, a flexible steel strip was used to seal the slot and it was decided to use a twin-cylinder configuration for improved efficiency. The resulting BS4/5 catapult would eventually be capable of accelerating a 50,000 lb aircraft to a speed of 105 knots, enough to cope with any naval aircraft then contemplated. Furthermore, the total installation was some 50 tons lighter than the BH.V and steam consumption was considerably less due to its more efficient direct action. Nevertheless, some of the figures are startling. The moving parts weighed around 10,000 lb and had to be brought to a standstill in only seven feet. In dispersing this amount of energy an instantaneous work rate of three million horse power was generated! As soon as it came to rest, the piston could be returned rapidly to its start position while a second aircraft was taxied into position. Here, retractable ramps acted as chocks under mainwheels, inset rollers allowed the aircraft to be positioned precisely on the catapult track, and a later refinement was the provision of retractable blast screens behind the aircraft to prevent damage to subsequent aircraft as engines were run up to full power. A well drilled catapult team could launch an aircraft every forty seconds from a single catapult and this rate could be halved if twin catapults were in operation.

The first sea trials of the steam catapult were carried out aboard the light fleet carrier HMS *Perseus*, which had been completed as an aircraft repair ship and did not normally have facilities for operating aircraft. During 1949/50 she was fitted with a steam catapult, which was installed below a raised platform on the forward section of the flight deck. For trials purposes aircraft were hoisted aboard by crane, being parked and prepared on the after section. Over the next two years a total of 1560 launches were successfully carried out, although 1,340 involved dummy loads or unmanned aircraft. The latter were drawn from stocks of surplus or time-expired airframes, which had their outer wing panels removed so that they would not sustain flight after the launch. At least that was the idea, but more than one of these old war horses showed a reluctance to come down once in the air and one Seafire actually started a climbing turn to starboard before diving back towards the carrier. Fortunately it ran out of fuel and crashed into the sea short of the carrier – much to the relief of all on board. Some of the dummy loads included a large water tank fitted with wheels and nicknamed 'Flying Flossie'. This had the advantage that it could be filled with varying amounts of water to simulate different aircraft weights and could be recovered for further trials, unlike the aircraft, which were strictly one shot.

The final proving trials involved a total of 166 manned aircraft and this went off without a hitch. Pilots reported a much smoother acceleration than was imparted by the conventional hydro-pneumatic catapults, which tended to impart a series of jerks as the cables took the strain of the launch. The experimental BXS.I steam catapult installed aboard HMS *Perseus* was capable of accelerating a 30,000 lb aircraft to a speed of 90 knots, but the later BS4 and 5 subsequently installed in British carriers were capable of launching 50,000 lb aircraft at speeds up 105 knots. The total of manned aircraft launches included a series made by US Navy aircraft when the ship visited the United States at the end of 1951 to demonstrate the new (and highly secret) technology. Some of the latest US Navy jets, including Panthers, Banshees and Skyknights, were successfully launched as well as older stalwarts such as the piston-engined Corsair and Skyraider. Once again, the trials were a resounding success and the US Navy immediately adopted the steam catapult for installation aboard its own ships.

The third major British contribution to jet carrier operations was a method to provide guidance to pilots making an approach to land. It was crucial that the pilot landed his aircraft centrally amongst the arrester wires at the correct speed, otherwise there was every chance of missing the wires or else over-stressing the arrester hook, which could shear off, leaving the aircraft careering down the deck. The traditional method of providing assistance was by means of a Deck Landing Officer (DLO) who would stand on the edge of the flight deck giving signals to the pilot by means of coloured or illuminated bats. A sequence of signals could indicate if a pilot was too high or too low, or whether correctly lined up. If all looked good the DLO would give the 'cut' signal at the appropriate moment and the aircraft would settle onto the deck and catch one of the wires. The DLO task required a considerable amount of skill (it was normally carried out by a practising pilot) and also a high degree of trust from the pilots concerned. One problem encountered by the Royal Navy towards the end of World War II when its carriers deployed to the Pacific was that the US Navy's signals were diametrically opposite to the British. Thus American raised bats meaning 'You are too low, Fly Up' would be interpreted as 'You are too high, Fly Down'. It took a little while and a few incidents before this one was ironed out (the Royal Navy adopted the US system). Also, the system was inevitably subject to human error, relying heavily on the skills of the individuals concerned, and it was difficult to execute and follow in poor weather conditions or at night.

A mechanical pilot-interpreted device would overcome a lot of these problems and a suitable installation was devised by Commander HCN Goodhart RN. Like many successful inventions, this one was based on a straightforward principle applied in a practical manner. In this case the approaching pilot would see a large mirror at the side of the flight deck as he approached to land on. In front of the mirror was a powerful spotlight whose beam was reflected towards the pilot. If approaching at the correct angle the pilot would see the reflection of the spotlight, which would be in line with fixed datum lights on either side of the mirror. If he went too high the blob of light (or 'meatball' as it was universally named) would appear to rise above the datum lights, or drop below if the pilot was too low. In addition, as the mirror was concave it actually caused the meatball to elongate vertically, forming a line of light with a bright centre, and this would move across the mirror depending on whether the pilot was correctly lined up or offset to the right or left. After successful shore trials, a Mirror Deck Landing Sight (MLDS) was installed aboard HMS *Illustrious* in October 1952 and over 100 landings were made without difficulty. The MLDS was adopted for general use and was fitted to all British carriers from 1954 onwards and, once again, was also taken up by the US Navy. In later years, this device was developed into the Deck Landing Projector Sight (DLPS), which operated on a similar principle but the mirror was replaced by a series of projected light beams that were variously visible to the pilot depending on his relative position. The DLPS was also fitted on a stabilised mounting, which gave a much more steady indication to the pilot when the ship was pitching and rolling. However, it was not introduced into service until 1959, after the period covered by this book.

As it was the Royal Navy that introduced these pioneering concepts, it is relevant to look at the British carriers that came into service in this period. It is a sad fact that no purpose-built conventional aircraft carrier was laid down in Britain after 1945 until the current CVFs (Future Aircraft Carriers), which will not enter service until around 2015. HMS *Invincible* of Falklands fame and her two sister ships were originally designed as helicopter-carrying cruisers and the addition of the STOVL (Short Take Off Vertical Landing) Sea Harrier was a fortuitous afterthought so they do not qualify as full-blooded carriers. As has already been recounted, the Royal Navy's carrier fleet in the early post-war years consisted of some of the Illustrious class fleet carriers and the Colossus class light fleet carriers. In the meantime work progressed slowly on two large carriers that were all that remained of original plans to build four Audacious class ships. The first pair, *Audacious* and *Ark Royal*, were laid down in 1942 and 1943, while work on *Eagle* commenced in April 1944 but the fourth ship (to have been named *Africa*) was never started. In January 1946 work on *Eagle* was halted and the contract cancelled, although the name was transferred to the first ship (*Audacious*), which was then launched as HMS *Eagle* in 1946. The name of course perpetuated the memory of the carrier sunk in the heroic Operation *Pedestal*, the largest of all the Malta convoy battles. *Ark Royal* spent a considerably longer period on the slipway and was not launched until 1950.

Although launched in 1946, HMS *Eagle* was not completed until October 1951. Despite this long time scale very few improvements were incorporated and she substantially conformed to the original design. The axial flight deck was just over 800 feet long and 112 feet wide. Two BH5 hydro-pneumatic catapults were installed forward and eight sets of double arrester wires aft. The catapults could launch 30,000 lb aircraft up to 75 knots while the wires could cope with similar-sized aircraft at a landing speed of 75 knots and up to six crash barriers were provided as well as roller positioners for the catapults. The traditional armament of sixteen 4.5 inch DP guns in eight twin turrets disposed quadrantly was followed, although the tops of the turrets were flush with the flight deck.

A secondary armament of no fewer than sixty-one 40 mm guns was carried, this including no fewer than eight sextuple 40 mm mountings, each controlled by a Mk.37 HA director. The main change over the original design was a much improved radar outfit, which included a long-range Type 960 with a novel Skyatron projected display system as well as a Type 982 fighter direction radar and Type 983 height finder radar.

HMS *Eagle* commissioned in March 1952 and she subsequently achieved a number of important milestones. Her original air group included 800 Squadron equipped with Attackers, the Royal Navy's first shipboard jet fighter squadron, and during her work-up period she was host to deck landing trials of several new types, including the Seahawk, Sea Venom and the turboprop Gannet. In 1953 she embarked the first operational Seahawk squadron (806 NAS) and also the first Skyraider AEW aircraft supplied under MDAP.

Despite being the Royal Navy's most modern carrier, HMS *Eagle* had been completed just before the full range of new British carrier aids had been tried and tested. However, her sister ship, HMS *Ark Royal*, had been considerably delayed on the building slips and was not launched until May 1950 when it was planned that she should commission in 1952. By this time the benefits of devices such as the angled deck and steam catapults were readily apparent and the decision was taken to delay her completion so that these improvements could be incorporated. Consequently she did not commission until February 1955, almost twelve years after her keel had been laid. Even then, the angled deck was only offset 5.5 degrees but this necessitated the suppression of the portside forward 4.5 inch gun turrets. A notable first for a British carrier was the inclusion of a deck edge lift on the port side amidships, although this was later removed in the course of a refit in 1959 as it was only of limited use, serving only the upper hangar deck. The ship's first air group, embarked in June 1955, comprised two Seahawk squadrons (800 and 898 NAS), one Gannet anti-submarine squadron (824 NAS) and a flight of Skyraider AEW aircraft (B Flight, 849 NAS) together with two Dragonfly and two Whirlwind helicopters for SAR and general duties. In 1956 she was involved in deck landing trials for the new Sea Vixen and

Scimitar before being taken in hand for further modifications in April. She did not recommission until November 1956, thus missing the Suez operation. Subsequently she saw considerable service, surviving until 1978 as Britain's last conventional aircraft carrier.

In the meantime, HMS *Eagle* was docked in mid 1954 for a nine-month refit in which she was equipped with an interim 5.5-degree angled deck, although the forward 4.5 inch guns were retained as well as the BH5 catapults. Recommissioning in February 1955, her air group included two Seahawk squadrons (802 and 804 NAS), two Wyvern strike squadrons (813 and 827 NAS) and the first operational deployment of the Gannet anti-submarine aircraft (826 NAS). *Eagle* subsequently took part in Operation *Musketeer* (Suez operations) but it was not until 1959 that she was paid off for a substantial modernisation refit, which added a fully angled deck and steam catapults as well as several other features. This refit took over five years and was not completed until 1964 – well outside the scope of this book. Suffice to say that the ship's full potential was finally realised and she gave excellent service until prematurely laid up in 1972.

Also entering service in the early 1950s were the three Albion class light fleet carriers, which represented a development of the Colossus/Majestic class. The main improvement was more powerful machinery, more than doubling output to 76,000 shp and raising maximum speed from 24 knots to almost 28 knots. Some armour plating was applied to vital areas such as the boiler and machinery rooms, and the magazines, while the flight deck was slightly enlarged to measure 733 feet by 103 feet. More significantly, hangar headroom was increased to 17.5 feet and a complement of fifty aircraft was anticipated. Original plans called for eight ships but in the event only four were laid down in 1944/5 and the remainder were cancelled after the war. Even work on those under construction was slowed down so that *Albion*, *Bulwark* and *Centaur* were not launched until 1947/8, after which they were laid up awaiting a decision on their future. The fourth ship, originally named *Elephant* but renamed *Hermes*, had an even more extended period in the slips and was not launched until 1953 and did not commission until 1959, having been under construction for some fifteen years!

Admittedly, when she did enter service she was equipped with a fully angled deck, steam catapults, a deck edge lift and a sophisticated radar system, but her subsequent career is outside the scope of this book.

The remaining three ships were eventually completed in 1953/4, their survival being in no small part due to the hangar headroom, which exceeded that of the Illustrious class then being retired. The original design had included four twin 4.5 inch DP gun turrets disposed quadrantly but these were deleted and the defensive armament was to comprise thirty-two 40 mm guns. A very comprehensive radar outfit was shipped, including a Type 960 search radar, two Type 982 fighter direction radars and a Type 983 height finder. The first to complete was HMS *Centaur*, which commissioned in September 1953 but after initial trials she was taken in hand at Portsmouth dockyard where an angled deck was installed. The work required to achieve this was kept to a minimum and involved a narrow extension to the port side of the flight deck amidships necessitating the removal of three twin 40 mm mountings. The deck centreline was offset 5.5 degrees and the arrester gear and barriers were re-aligned to conform. By the autumn of 1954 the work was complete and *Centaur* deployed to the Mediterranean with an initial air group comprising piston-engined Sea Furies and ASW Avengers. It was not until the following year that jets came aboard in the shape of Seahawks of 805 and 806 Squadrons.

Completion of the other two ships, *Albion* and *Bulwark*, was delayed so that the angled deck could be installed. When *Albion* commissioned in May 1954 it was claimed that she was the first new-build carrier in the world to be so equipped. In this case the deck was angled at 5.75 degrees and she was also the first to be fitted with the mirror landing sight. *Albion* subsequently deployed to the Mediterranean with an air group made up of Seahawk jet fighters, Wyvern turboprop strike aircraft and piston-engined Skyraiders for AEW. *Bulwark* commissioned in November 1954 and initially replaced *Illustrious* as a trials and training ship before embarking a unique air group that consisted entirely of Seahawk jet fighters (801 and 898 NAS) backed up by Skyraider AEW aircraft and Dragonfly SAR helicopters. Both *Albion* and *Bulwark* were present in the Suez operations

(November 1956) but at the end of the decade both were converted to helicopter-carrying assault ships as they were considered too small to effectively operate the new generation of naval aircraft then coming into service.

Both *Albion* and *Bulwark* retained their BH5 hydro-pneumatic catapults while operating fixed wing aircraft but *Centaur* was modified during a refit lasting two years from September 1956 in which steam catapults were fitted. Although she subsequently operated Scimitars and Sea Vixens, the numbers that could be embarked were limited and *Centaur* was eventually retired in 1965.

Thus by 1955 the effective Royal Navy carrier strength consisted of the two large fleet carriers, *Ark Royal* and *Eagle*, and the three Albion class light fleet carriers. In addition, a few of the earlier Colossus class were retained for a while in secondary roles, although one of these, HMS *Warrior*, was fitted with an angled deck while under refit in 1955/6. In order to boost this force it had been proposed to modernise some of the Illustrious class but in the event only HMS *Victorious* was taken in hand. *Illustrious* and *Formidable* were also considered for modernisation but with the limited funds available the Admiralty decided to concentrate on the newer ships then under construction and, in any case, a survey of *Formidable* revealed that she was in poor condition, partly due to wartime damage that had been hastily repaired at the time. However, work progressed on *Victorious* and from March 1950 she became a semi-permanent fixture at Portsmouth dockyard. Some major alterations were made, including cutting the ship in half so that she could be lengthened by 30 feet while her beam was increased by 8 feet. Eventually new machinery was installed and virtually every compartment was stripped and refitted. Originally it was intended that the ship would be able to operate an air group of fifty-four postwar jet aircraft from a conventional axial flight deck. However, the development of the angled deck and steam catapults resulted in the design being modified to accommodate these features and it was not until 1958 that *Victorious* finally commissioned. At that time she was the only Royal Navy carrier with a full angled deck, it being canted 8.75 degrees off the centreline. She was also the first to be fitted with the new Type 984 '3D' radar and all in all she was a very valuable addition to the fleet, although her subsequent career is outside the scope of this book.

In contrast to the Royal Navy, the US Navy not only retained the bulk of the wartime Essex class carriers, but was able to constantly modify and upgrade them so that many remained in frontline service well into the 1970s. The upgrading of some ships under the SCB-27A programme has already been covered. In 1949 further conversions designated SCB-27B were proposed but shortage of funds (even the mighty US Navy suffered from financial constraints but nowhere near as badly as the British with their depressed post-war economy) prevented their implementation. However, by 1951 it was becoming apparent that a further upgrade was still required if these ships were to be able to operate some of the newer types likely to enter service such as the McDonnnell F3H Demon. One major change was to be the installation of new C-10 catapults using explosive powder charges, replacing the H-8 hydraulic catapults. Mk.VII arrester gear capable of handling heavier aircraft would replace the earlier Mk.V wires and jet-blast deflectors behind the catapults were to be fitted for the first time. The deck was strengthened to accommodate parked aircraft weighing 60,000 lb and an additional deck edge lift was installed on the starboard side aft of the island, replacing the previous after centreline lift, and this arrangement would considerably speed up deck operations. All these improvements added weight and the full load displacement rose to 42,000 tons, necessitating an increase in the hull bulges, which, while maintaining stability at acceptable levels, reduced speed to just over 30 knots.

One novel feature of the SCB-27C project was the manner in which jet fuel capacity was substantially increased. It had been found that mixing aviation gasoline (Avgas) as used by the piston-engined aircraft with the ship's own 'heavy-end' bunker oil produced a usable jet fuel. As the bunker oil was non-volatile it could be stowed in normal tanks without the extensive protection required for the more volatile aviation fuels. Using this method up to 739,000 US gallons of jet fuel could be made available, although a knock-on effect was a reduction in the ship's cruising range from 11,500 nm to around

8,500 nm at 20 knots. However, given the extensive tanker support available to the US Navy, this was not a serious drawback. Even while the SCB-27C scheme was under consideration, the US Navy learnt of British developments such as the steam catapult and the angled deck. Although the angled deck had very obvious advantages and was quickly adopted, the steam catapult was actually something of a life saver for the US Navy as the C-10 catapult had proved unsuccessful and could possibly have caused the abandonment of the SCB-27C project if the British alternative had not been available. Steam catapults were subsequently manufactured under licence in the United States, initial versions being designated C-11, and these were installed in the SCB-27C ships.

As far as the angled deck was concerned (initially termed 'canted deck' in US Navy parlance), the Americans were keen to try out the idea in practice as soon as possible and the USS *Antietam* was selected for the prototype installation. This ship had been completed in November 1945 but had been laid up in reserve in 1949 before being re-activated for Korean War service in 1951. In May 1952 she was at New York Navy Yard where a 10.5-degree angled deck was fitted, the work being completed in April 1953. She was immediately involved in extensive trials to test the new concept and these were an outstanding success. As part of the trials programme she visited the UK in June 1953 for a week of operations with current British jet aircraft such as the Seahawk and Attacker. Subsequently *Antietam* remained in service until 1963 but she was not further modified.

However, the introduction of the angled deck into the American ships resulted in another modification programme designated SCB-125. While the main feature was the new deck layout a number of other improvements were incorporated. Up to this point all Essex class carriers had featured an open bow on which were mounted several 40 mm AA guns. However, this arrangement was prone to damage in severe sea conditions so the SCB-125 programme introduced a fully enclosed bow, known as a 'hurricane' bow, which was remarkably similar to the profile adopted by all British carriers from before World War II. Other features that assisted flight deck operations included improved lighting and strengthened crash barriers, the latter still being retained for emergency use, although they were not required in normal flying operations. The number of arrester wires was halved due to the fact that with the angled deck a pilot had the option of a go around, or 'bolter', if he missed the wires.

The application of the various modification programmes to the ships of the Essex class inevitably resulted in several variations. Between 1948 and 1953 eight ships (*Essex, Yorktown, Hornet, Randolph, Wasp, Bennington, Kearsarge,* and *Lake Champlain*) all underwent the basic SCB-27A programme, while the USS *Oriskany* was actually completed to this configuration in September 1950. Another three ships (*Intrepid, Ticonderoga* and *Hancock*) were modified as SCB-27C between 1951 and 1954 but then subsequently received the SCB-125 modifications, including the angled deck in further refits between 1955 and 1957. Three more ships received the full SCB-27C and SCB-125 modifications in a single refit between 1951 and 1955, these being *Lexington, Bon Homme Richard* and *Shangri La*. The *Oriskany*, originally completed as an SCB-27A, was modernised between 1957 and 1959 under a scheme designated SCB-125A, which included the laying of a metal flight deck in contrast to the traditional wooden deck surface on all other Essex class ships. Three ships received no modifications in their careers as conventional fixed wing aircraft carriers, these being *Boxer, Princeton* and *Valley Forge*, although between 1959 and 1961 they were converted to helicopter assault ships (LPH) along the lines of the successful British conversions of the smaller *Albion* and *Bulwark*. Finally, five ships (*Franklin, Bunker Hill, Leyte, Tarawa* and *Philippine Sea*) were not modernised at all during their post-war careers, the first two having received severe damage during World War II, although both were repaired.

The Essex class carriers with the SCB-125 modifications remained in service up to the time of the Vietnam War and proved capable of operating the new combat aircraft that were entering service or under development in the early 1950s (and are described in the next chapter). These included Cougar, Demon and Crusader jet fighters, and the Skyhawk and Skywarrior attack bombers. However, those that were involved in the Korean War, although operating the first generation jets such as the Panther and Banshee, were mostly the later ships that remained unmodified. The first angled deck ship to deploy with an operational air group was the USS *Shangri La*, which recommissioned in January 1955, well after the end of hostilities. In summary, in the post-war era the US Navy reaped the benefit of a large wartime construction programme and was provided with a carrier force that proved capable of operating the newer, heavier and larger aircraft that came into service in the decade after 1945. Partly this was because the US Navy had always prioritised on the aviation requirements when designing aircraft carriers, in contrast to the Royal Navy in which the compromises required between the demands of the ship and its air group were not always decided in favour of the aviation interests. Consequently, as aircraft grew in size and performance the basic requirements such as flight deck size, lift numbers and capacities and hangar stowage were already in place.

Also, of course, the US Navy possessed the three large Midway class carriers, which were much bigger than the Essex class and had no difficulty in accommodating new aircraft. Consequently, many of the early jet operations took place aboard these ships. Nevertheless, even these great ships required updating to meet new challenges and in 1947/8 all three had their flight decks strengthened to allow them to operate the AJ-1 Savage nuclear bomber, which was then undergoing flight testing prior to its operational debut in 1950. At around that time the Midway class was also fitted with facilities to allow it to launch Regulus 1 missiles. These were an early version of the cruise missile concept and after a rocket-boosted launch from a platform erected on the carrier flight deck, they used a jet engine to give the missile a range of around 200 miles carrying a 120-kiloton nuclear warhead. Although the launcher was a simple structure, ship modifications required spaces for missile stowage and assembly, as well as secure magazines for the nuclear warheads.

As with the Essex class, the three Midways underwent modernisation programmes, of which the first was designated SCB-110. The *Franklin D Roosevelt* was docked from May 1954 to April 1956 and *Midway* followed in September 1955, her conversion taking almost exactly two years. The SCB-110 programme included the construction of an angled deck offset at eight degrees and having a length of 482 feet. Two C-11 steam catapults were fitted and a hurricane bow profile was adopted. The lift arrangements were considerably altered, the after centreline lift was removed and replaced by a new deck edge lift on the starboard side, while a second edge lift on the port side was positioned at the forward end of the angled deck. New arrester gear with a reduced number of wires, as in the Essex refits, was also installed and aircraft fuel stowage considerably increased. All this added considerable weight and as partial compensation some of the armament was removed leaving a total of only ten 5 inch/54 cal and twenty-two 3 inch/5 cal AA guns. The aircraft capacity was now around eighty and the squadrons embarking were to be equipped with the new range of naval combat aircraft, including the F3H Demon, F8U Crusader, A3D Skywarrior and A4D Skyhawk, as well as the ubiquitous AD Skyraider.

The third ship, USS *Coral Sea*, did not undergo modernisation until 1957–60, outside the scope of this book. However, it is of interest to note that her SCB-110A programme involved the removal of all centreline lift lifts, which were replaced by one deck edge lift on the port side amidships, and two on the starboard side situated forward and aft of the island superstructure. An additional C-11 steam catapult was installed along the angled deck to supplement two bow catapults. Between 1966 and 1970, *Midway* underwent a further modernisation (SCB-101.66) but cost issues resulted in plans for the *Franklin D Roosevelt* to be similarly modified being dropped. Because of this she was the first of the class to be paid off, in December 1977, but the others had much longer careers. *Coral Sea* lasted until 1990 and *Midway* took part in the first Gulf War in 1991 before being retired the following year after forty-seven years' service – a remarkable achievement, especially when it is considered that the British *Ark Royal* and *Eagle*, laid down at the same time, only achieved twenty-three and twenty-one years in service. In fairness it should be pointed out that this was more due to differing political outlooks rather than any inherent problems with the ships themselves.

Today the US Navy operates a fleet of nuclear-powered super carriers and all the ships of World War II origin have long since been retired, mostly cut up for scrap. However, the concept of the super carrier had its origins as far back as late 1945 when opinion in the US Navy was moving towards a requirement for a naval bomber that could weigh as much as 100,000 lb and be capable of carrying a 12,000 lb bomb over a radius of 2,000 miles. This was almost in the same class as the Boeing B-29, the Air Force's standard bomber, and it would obviously require a very large ship from which to operate. The concept of a pure bomber with a strategic capability was new to the US Navy. Previously it had concentrated on weapons systems such as torpedoes and conventional bombs that could sink ships, and had no requirement for anything else. However, in the post-war era naval air power was much more likely to be deployed against land targets and the US Navy became embroiled in a struggle with the United States Air Force (formerly the US Army Air Corps until 1947) over supremacy in the strategic nuclear strike role in which the Navy was determined not to lose out.

A sketch bomber design known as ADR42 had a wingspan of 116 feet, reducing to 44 feet with wings folded, and the US Navy wanted a ship that could carry and operate twenty-four of these aircraft. The result was a very large aircraft carrier with an overall length of 1,090 feet, a hull beam of 130 feet and an overall beam of 190 feet measured over the flight deck and projecting sponsons. The standard displacement was estimated at 66,850 tons, rising to 78,500 at full load. Despite the sheer size, the ship would be capable of 33 knots with a 280,000-shp four-shaft machinery installation. Defensive armament would comprise eight single 5 inch/54 cal and six twin 3 inch/50 cal AA guns. However, the most interesting feature was the flight deck, which was entirely flush with no superstructure, the ship being conned from a small platform forward on the starboard side, and smoke from the boilers being vented through outtakes on either side of the after end of the flight deck, rather in the manner common to many Japanese carriers in World War II. In order to launch a strike in a reasonably quick time frame, four catapults were envisaged, two in the bow and one on each side amidships angled outboard. The latter arrangement was almost a forbear of the angled deck concept but at the time no one thought of its application for landing aircraft. Although the original expectation was that the bombers would have to be stowed in a deck park, the sheer size of the ship allowed a large hangar with a 28-foot headroom – greater than subsequent carriers, which only have 25 feet. By 1947 when the design was being refined the draft characteristics showed a revised air group of only twelve ADR42 bombers but up to fifty-four F2H Banshee jet fighters for escort purposes. Alternatively up to eighteen smaller ADR45A attack aircraft could be accommodated. The latter were quoted as having an all-up weight of 57,000 lb and were to be powered by four turboprops giving a radius of 750 miles. Allocated the hull number CVA58 and named USS *United States*, the ship was laid down on 19 April 1949 at Newport News. In the meantime the power struggle between the Air Force and the US Navy reached a climax with the former victorious, securing funds for a strategic bomber force so that work on the carrier ceased after only four days and the contract was cancelled.

However, the work put into the design of the flush-decked carrier was not all lost and in 1950 studies began for a new slightly smaller 60,000-ton carrier. The outbreak of the Korean War released funding for the project, which was based around the Douglas A3D Skywarrior jet bomber that began development in 1949 and subsequently flew in 1952. The flush-deck layout was initially retained and still featured when the first of a class of four ships, the USS *Forrestal* (CVA59), was laid down in July 1952. The four-catapult layout was also retained but these were the C-10 powder charge type, which were to prove unsatisfactory in operation, quite apart from the requirement to provide additional magazine space to stow the large number of charges required. Although the catapults were canted out to port and starboard, landing aircraft were still intended to land along the deck axis. However, the adoption of the British concept of an angled deck permitted a conventional island superstructure to be built up on the starboard side of the flight deck, the catapult on that side being moved across to the port side where one catapult was almost parallel to the ship's axis and the other was aligned with the 8-degree offset angled deck. This arrangement permitted two aircraft to be positioned simultaneously on the portside catapults before being launched in turn. Again, the adoption of the steam catapult was a godsend and solved many problems that would have arisen if the powder charge type had been fitted.

As completed in October 1955, the USS *Forrestal* displaced 60,000 tons in standard trim but this rose to 78,000 tons at full load. At that time she was the largest warship ever built and could carry around ninety aircraft. Her first air group comprised three fighter squadrons equipped with F9F Cougars or F2H Banshees, two attack squadrons with A4D Skyhawks and the indestructible AD Skyraider, and one heavy attack squadron with the A3D Skywarrior twin jet bomber. In addition there were small numbers of SAR helicopters and AEW aircraft as well as a photo reconnaissance detachment. Although the original Forrestal class numbered only four ships (all in service by 1959), they were followed by four more Improved Forrestal class and although there were some significant differences among the individual ships, they all followed the same basic design. Indeed, the first nuclear-powered carrier, USS *Enterprise* (CVAN65), differed little in overall layout and the subsequent Nimitz class again conformed, although slightly larger. In effect therefore, the optimum carrier size and layout for the operation of modern naval jet aircraft had been established by 1955. Since then there have been few major changes other than the adoption of nuclear power. Even the latest US carriers laid down in the twenty-first century will retain the overall configuration, although the ships' systems based around new electronics and massive computing power offer a Control, Command and Communication system that is light years removed from that available in 1955. One other change is that the long-serving steam catapult will give way to an Electro-Magnetic Aircraft Launching System (EMALS), which is currently under development but shows great promise.

6

SWEPT WINGS AND
SUPERSONICS

By the outbreak of the Korean War in 1950, most of the jet aircraft described in Chapter 4 had entered service or were undergoing development, and the design of their potential successors was already well underway. This new generation would have the benefit of more powerful engines and would be able to take advantage of some major aerodynamic advances.

In the case of the Royal Navy, an incremental approach was adopted, which led to a series of interesting prototypes but no production orders. This was partly due to financial constraints but also because policy was geared to the idea that any major conflict would not break out until 1957 and so aircraft development was concentrated on the most advanced types that could be available by that date. The incremental approach is clearly illustrated by a series of prototypes developed from the Royal Navy's first jet fighters, the Attacker and the Seahawk, although in the event neither resulted in an operational carrier borne fighter. Design work on the new aircraft began as a result of specifications issued by the Air Staff in 1946. Of these, E.41/46 called for a swept wing development of the Attacker and this resulted in the Supermarine Type 510, which flew on 29 December 1948. A major change was that all the flying surfaces (wings, tailplane, fin) were swept back at forty degrees but in most other respects the basic Attacker configuration was retained, including the tailwheel undercarriage. The same Rolls-Royce Nene jet engine was also installed but, even so, a considerable increase in performance was soon demonstrated, including a top speed in excess of 630 mph at medium altitudes and a limiting Mach number of 0.93. In general the handling characteristics received favourable comment except for longitudinal instability near the stall, which affected manoeuvrability at high altitudes where aircraft were operating much closer to their stalling IAS due to the much thinner air.

The Royal Navy closely monitored progress with the Type 510 and the prototype was modified for naval trials leading to the first deck landing by a swept wing aircraft on 8 November 1950, aboard HMS *Illustrious*. For these trials the aircraft was fitted with Rocket Assisted Take Off Gear (RATOG) and a total of fifteen landings and rocket assisted take-offs were successfully completed. On the last take-off some of the rockets failed to ignite and the aircraft yawed dangerously, causing one wingtip to strike the top of a gun turret before the pilot was able to recover and fly the aircraft ashore. Subsequently the aircraft returned to the RAE at Farnborough for further high-speed flight tests and the naval equipment was removed. Despite the take-off incident, the deck landing trials had gone well but the Admiralty took no further interest in this particular line of development. Fitted with an all-flying tailplane and a tricycle undercarriage, the second prototype became the Supermarine Type 525, which formed the basis for the Swift that entered service with the RAF in 1954, although persisting problems with high-altitude manoeuvrability limited its usefulness.

In an almost parallel scenario, Hawker also evolved and flew a swept wing development of the P.1040 Seahawk. The origins of the swept wing project actually went back to October 1945 when Hawker proposed such a project, although powered by a rocket engine, perhaps inspired by the German Messerschmitt Me.163 interceptor. Although this was not followed through in its original form it did lead to the issue of Specification E.38/46, which resulted in an order for two prototype Hawker P.1052s. This was very similar to the basic Seahawk design with the single exception that wings swept thirty-five degrees at quarter chord was fitted, although the original tail surfaces with a straight tailplane were retained. The first flight was on 19 November 1948 by which time serious consideration had been given to placing a production order as the promised performance was expected to be well in advance of that of the Seahawk or the RAF's Meteor. Had that been done, swept wing

Seahawks could have entered service during the Korean War period but no order was forthcoming as it was decided to wait for other more advanced aircraft to become available. Although the Royal Navy was eager to assess the P.1052 aboard a carrier, such trials were delayed due to accidents in the flight test programme. It was therefore not until May 1952 that the first prototype (VX272), modified for naval use and painted in Royal Navy colours, made its first landing aboard the new carrier HMS *Eagle*. The following month it was fitted with a new swept tailplane and carried out high-speed testing with the RAE before being grounded after a forced landing in the September. Subsequently the second prototype (VX279) was rebuilt with a straight-through jet exhaust (instead of the bifurcated exhaust) and a swept fin and tailplane. This was designated P.1081 and flew in June 1950 but was lost in an accident the following year. By this time Hawker had moved ahead with the Avon-powered Type 1067, which became the Hunter and was ordered in quantity by the RAF although at the time no naval development was ever considered. As a footnote, Hunter single-seat fighters did eventually serve with the Royal Navy in the 1960s in the shore-based operational training role.

Although both the Supermarine Type 510 and the Hawker P.1052 provided valuable data on the operation of swept wing aircraft aboard carriers, as well as offering a significant increase in performance, the Admiralty did not consider placing production orders as there were more advanced aircraft already under development. One of these was from the Supermarine stable but was to have an extremely convoluted career before eventually entering frontline service as the Scimitar in 1958. Interestingly the genesis of this aircraft went back to the flexible deck concept described in Chapter 4 and in response to Admiralty requirements a twin-engined straight winged design was evolved as early as 1945. This was the Type 505 in which two Rolls-Royce AJ65 turbojets (an early version of the Avon series) were mounted side by side in the central fuselage and an unusual Vee tail arrangement was adopted to keep the tail surfaces clear of the jet exhaust. Although a conventional straight wing was used, it was very thin and had a thickness chord ratio of only 7 per cent. This was one of the benefits of dispensing with the undercarriage, which would have required a thicker wing to accommodate it.

By 1947, despite the fact that the flexible deck trials had not yet started, the Admiralty decided that the Supermarine design should be modified to include a tricycle undercarriage. This involved a new wing with thickness/chord increased to 9 per cent and also an increased wing span to cope with revised low-speed handling requirements. In this form the aircraft flew on 31 August 1951 and subsequently carried out deck landing trials aboard HMS *Eagle* in May 1952. However, by that time Supermarine had gained experience of swept wing aircraft with the Type 510 modified Attacker and decided to proceed with a swept wing version of the Type 508, which then became the Type 525. The adoption of a swept wing necessitated many changes, including a taller undercarriage with a wider track due to the geometry of the swept wings. Extensive conventional high lift devices such as double-slotted flaps and leading edge slats were incorporated but subsequently the idea of 'blown flap' was developed. In this system air was bled off from the engine compressor stages and was fed out of a narrow slot in the wings just ahead of the trailing edge flaps. This stream of energised air prevented a breakdown in the airflow over the flap (retaining the laminar flow effect beyond the point where it would normally break away and cause loss of lift) and the result was a substantial reduction in stalling speed and angle of attack while on final approach to land. The speed reduction was in the order of 15 knots and this was a highly desirable benefit for carrier landings.

The Admiralty was sufficiently interested in the Type 525 to issue a draft Specification N.113D around the proposals and ordered two (later increased to three) prototypes in 1951. The first of these (VX138 – originally ordered as the third Type 508) flew in April 1954. While the performance was generally up to expectations, there were numerous problems associated with the handling of the aircraft at high speed and these resulted in a number of modifications that were incorporated in the production examples, which became the Type 544 Scimitar. The most obvious external feature was a change from 10 degrees dihedral to 10 degrees anhedral on the swept tailplane. Other

aerodynamic improvements included a saw tooth leading edge
and re-profiled wingtips. Less obvious was the inclusion of
powered flying control surfaces with fully duplicated systems.

The N.113D Scimitar was originally intended as a swept
wing successor to the Seahawk but while under development
its role was changed from being a pure single-seat fighter
to include the ability to undertake low-level strike missions,
armed with nuclear bombs if necessary. Consequently the airframe needed to be extremely strong to cope with the stresses
associated with operating at high speed in turbulent air at low
level and several innovative manufacturing techniques were
introduced. The profile of the fuselage was altered to comply
with the new concept of area rule and the lateral air intakes
were enlarged to cope with the demands of the latest Rolls-
Royce Avons, which would be initially rated at 10,000 lb thrust.
The first true Type 544 Scimitar, fully representative of the
operational aircraft, flew on 19 January 1956. Subsequently,
production examples reached the Royal Navy early in 1958 and
the first operational squadron, 803 NAS, embarked in HMS
Victorious in the following September. At the time of its introduction into service the Scimitar was the fastest, heaviest and
most powerful aircraft ever to enter service with the Royal
Navy. Armed with four 30 mm cannon as a fighter, it could also
carry up four 1,000 lb bombs and subsequently was also armed
with Bullpup air-to-surface or Sidewinder air-to-air missiles.
On paper it was a very potent interceptor with an initial rate of
climb in excess of 12,000 feet/min but its handling at high
altitudes was unsatisfactory (a feature it shared with its stablemate the Swift) and its true forte turned out to be as a low-level
strike aircraft with a nuclear capability. Despite the high thrust
available and the area ruled fuselage, the Scimitar remained
stubbornly subsonic except in a dive, leading one American
pilot to comment, 'Jeez, only you Brits could build an aircraft with so much power which won't go supersonic!' The
Scimitar's operational career falls outside the time scale of the
book but it does earn a place as the Royal Navy's first swept
wing fighter.

The Scimitar was always intended as the replacement for the
Seahawk and the latter was usually paired aboard carriers with

the de Havilland Sea Venom all-weather fighter. It was natural therefore that the de Havilland company would be interested in providing a successor for the Sea Venom, although the initial project proposed in 1946 was also put forward to meet an RAF requirement for an advanced night fighter and it was the Air Ministry that ordered two prototypes in 1948. The resulting de Havilland DH.110 followed the twin boom layout of the Vampire and Venom series but introduced a sharply swept wing based on that used in the DH.108, the first British aircraft to exceed the speed of sound. However, the DH.110 was much larger than its predecessors and was powered by two Rolls-Royce Avon turbojets in the rear of the fuselage nacelle. Unusually, the pilot's cockpit was offset to port to allow space, popularly known as the 'coal hole', for the radar operator and his equipment in the lower starboard side. Powered controls were fitted and the tailplane rode high up between the swept fins so as to be well clear of the jet efflux. An armament of four 30 mm cannon was proposed. The DH.110 prototype (WG240) flew in September 1951 and in April the following year demonstrated a transonic capability by exceeding the speed of sound in a shallow dive. A second prototype flew in July 1952 but was destroyed in a horrific accident when it broke up in the air while participating in the Farnborough Air Display. Pieces of wreckage, including the engines, scythed through the crowd, killing several spectators as well as the pilot, John Derry, and his observer.

This accident was a severe setback to the programme and the remaining aircraft was grounded for extensive modifications, including an all-flying tailplane and wing leading edge extensions outboard of the wing fences. By the time it was flying again in June 1954 the RAF had decided against placing any orders, preferring the delta-winged Gloster Javelin instead. However, the Royal Navy was now waking up to the performance potential of the DH.110 and the design was substantially modified to meet naval requirements, the name Sea Vixen being subsequently adopted. Up to this time the Royal Navy had been actively pursuing a project to produce a modified Sea Venom that would have had a new thin section swept wing and this had got to the stage where two prototypes had been ordered, although these were now cancelled in favour of the twin-engined DH.110. Changes to the new aircraft included the provision of an arrester hook, power folding wings, hydraulic nosewheel steering and a long stroke undercarriage for carrier landings. A major change was in the armament, the 30 mm guns being removed in favour of an all-missile armament. The space previously occupied by the guns was used to install retractable housings for a total of twenty-four 2 inch unguided air-to-air rockets, while up to four de Havilland Firestreak air-to-air missiles with an infra-red homing system would be carried on underwing pylons. For the strike role, the missiles could be replaced with four 500 lb or two 1,000 lb bombs, or a variety of air-to-air or air-to-surface rockets. All in all, the Sea Vixen was to be one of the most capable aircraft ever to serve with the Fleet Air Arm.

The first pre-production naval aircraft flew on 20 June 1955, by which time orders had already been placed for forty-five Sea Vixen FAW.1s and ultimately 114 were built. Later developments included the FAW.2 with the booms extended forward over the wings to give additional fuel tankage and fifteen of these were built while a further sixty-seven were produced by converting some of the earlier aircraft. In time scale the Sea Vixen was almost two years behind the Scimitar with the first operational squadron forming in July 1959 and embarking in HMS *Ark Royal* in March 1960. Thus by the end of the 1950s the Royal Navy was at last introducing swept wing transonic fighters into frontline service and had two powerful and capable aircraft to form the backbone of the carrier air groups. Both gave excellent service in the following decade. However, during the 1950s the US Navy had introduced no fewer than five advanced swept wing fighters and by 1960 the first of the supersonic Mach 2 world record beating McDonnell F4H Phantoms were being delivered. Indeed, they had even experimented with vertical take-off fighters and a transonic seaplane jet fighter. Bearing in mind the original British lead in jet propulsion immediately after World War II, it is instructive to see how the Americans began to forge ahead so quickly.

In fact, the US Navy was inexplicably slow off the mark to apply the principle of swept wing aerodynamics and its arch rival, the USAF, initially made all the running. This is even more surprising when it is realised that the fighter that effectively gave them command of the skies in the Korean War was actually developed from a naval jet fighter. This was the famous North American F-86 Sabre, which had its origins in the FJ-1 Fury that had been ordered by the Navy in 1944 (see Chapter 4). This was a straight-winged single-engined fighter characterised by its then unique nose intake and straight-through jet configuration. The USAAF ordered a land-based version under the designation XP-86 but a courageous decision was taken to delay delivery and production for a year so that the design could be recast to incorporate a 35-degree swept wing and powered flying controls. Even so, the first XP-86 flew as early as 1 October 1947 and a few months later became the first US fighter to exceed the speed of sound in a shallow dive. The initial production version became the F-86A Sabre and by the end of 1949 two USAF fighter groups were equipped with the new fighter, which proved to have excellent handling characteristics. Subsequently several thousand were built in the United States with licence production being undertaken in Canada, Italy, Australia and Japan. The Sabre can be regarded as one of the most famous aircraft ever built.

Despite its naval origins, the US Navy did not take a serious interest in the Sabre until after the outbreak of the Korean War and eventually ordered three prototypes of a naval version in March 1951. These were designated FJ-2 Fury and eventually some 200 were produced. In most respects they were standard F-86E Sabres powered by 6,000 lb thrust General Electric J47-GE-2 turbojets, which gave a maximum speed of 676 mph at sea level, an initial rate of climb of 7,250 feet/min and a combat ceiling of 41,700 feet. The only modifications for naval service were the obvious ones of an arrester hook, catapult attachment point and power folding wings, as well as a lengthened nose-wheel oleo to increase the angle of attack for catapult launches. In addition the armament was changed from six 0.5 inch machine-guns to the standard US Navy fit of four 20 mm cannon. Despite these limited changes, FJ-2 Furies did not begin to reach operational units until January 1954 due to the priority accorded to Sabres for the Air Force and most were allocated to

USMC squadrons. Later that year US Navy squadrons began to receive a more advanced version of the Fury under the designation FJ-3. Development of this had started in March 1952 and the main change was the installation of a much more powerful Wright J65-W-2 engine rated at 7,800 lb thrust. This necessitated an enlarged nose intake and a slightly deeper fuselage profile and production aircraft were powered by the slightly derated J65-W-4 engine. In passing it should be noted that the J65 was in fact a licence-built version of the British Armstrong Siddeley Sapphire turbojet and this was used by several other US Navy jets.

Some 538 FJ-3s were built between 1953 and August 1956 and the aircraft equipped no fewer than seventeen US Navy and four Marine squadrons. The first unit was VF-173, which received its Furies in September 1954 and was deployed aboard the USS *Bennington* in May 1954. An aircraft from another US Navy fighter squadron (VF-21) was the first jet to land aboard the new carrier USS *Forrestal*, this event occurring on 4 January 1956. In that year also missile-equipped FJ-3Ms began reaching the fleet, these being fitted to carry up to four Sidewinder air-to-air missiles. The final Fury variant was the FJ-4, which first flew in October 1954 and represented a complete redesign with the objective of increasing range and endurance. The requirement to carry 50 per cent more fuel resulted in a new fuselage out-line while thinner wings and tail surfaces were also fitted. To improve handling aboard carriers, a wider track undercarriage was fitted. A total of 152 FJ-4s were produced but these were used almost exclusively by Marine squadrons, replacing the earlier FJ-2. Deliveries began in February 1955. These were intended for use in the close support role and the four under-wing pylons could carry either bombs or missiles. The ultimate attack version was the FJ-4B, which did not appear until the end of 1956 but this had six weapons stations and was fitted with a Low Altitude Bombing System (LABS), which enabled the Fury to use the toss bombing technique to deliver a tactical nuclear weapons. The FJ-4B was issued to nine US Navy and three Marine attack squadrons and when production ended in May 1958 a total of 1,112 swept wing Furies had been delivered, making it one of the most significant naval fighters of the period.

To some extent the development of a naval version of the F-86 Sabre was driven by the sudden appearance of the swept wing MiG-15 over Korea and the realisation that the US Navy's current straight-winged jets were outclassed. The same motive also led directly to a swept wing development of the existing Grumman F9F-5 Panther. In fact, Grumman had investigated the possibility of swept wing variant when the original Panther design had been proposed but although some design work was done, the decision was made to concentrate on getting the Panther into service. There were also doubts about the suitability of swept wing aircraft for carrier operations due to the problem experienced at that time with low-speed handling. However, the appearance of the Russian-built MiG swept such concerns aside and Grumman was authorised to proceed with the construction of three swept wing Panthers in December 1950. The project was given the highest priority with the result that the prototype F9F-6 Cougar flew in September 1951 and VF-32 received the first production examples in November 1952, although by the time the aircraft was ready for operational deployment the Korean War had ended.

Compared with the Panther, the most obvious change was fitting a new wing with 35 degrees of sweepback at quarter chord, as well as a similarly swept tailplane, although the vertical tail surfaces were substantially unchanged. Inevitably the stalling and approach speeds increased but to assist low-speed handling the chord of the leading edge slats and trailing edge flaps was increased, larger flaps were fitted below the centre section, and rudder controls were boosted by the addition of a yaw damper. The forward fuselage was lengthened by 2 feet and the wing centre section, which included the air intakes, was also extended forward. The lengthened fuselage allowed internal fuel capacity to be increased to allow for the fact that wingtip tanks (as on the Panther) could not be fitted. Flight testing resulted in changes, including the adoption of an all-flying tail and the introduction of spoilers to replace conventional ailerons and improve lateral control. Finally, it was found necessary to fit conspicuous wing fences to reduce

Pilot's view landing aboard the carrier HMS *Centaur* which has been modified to incorporate an angled deck and has been fitted with the Deck Landing Mirror Sight prominent on the port side. Note the clear deck available ahead for a 'bolter' if none of the arrester wires are trapped. (*Fleet Air Arm Museum*)

A close up view of a Deck Landing Mirror Sight (DLMS) installed ashore at RNAS Brawdy for training purposes. Reflected in the mirror is the reference light installed further along the runway and Seahawk about to land. (*Fleet Air Arm Museum*)

The ultimate development of the DLMS was the fully stabilised Deck Landing Projector Sight. As can be seen this was a much more sophisticated and complex piece of equipment. (*Fleet Air Arm Museum*)

The completion of HMS *Ark Royal* was delayed for almost two years so that she could be fitted with an angled deck. When finally commissioned in early 1955, she was also the first British carrier to incorporate a deck edge lift. (*Author's Collection*)

In the post-war era the US Navy was in the fortunate position of having numerous Essex class carriers available. This is USS *Leyte* (CV.32) which commissioned in April 1946 and is substantially unaltered in this February 1949 photo. The carrier fleet was augmented by a fleet train of auxiliary vessels which ensured that the carrier groups could remain at sea for long periods. (*US National Archives*)

The USS *Shangri La* (CV.38) was an Essex class carrier completed in September 1944. Between 1952
and 1955 she was modernised under the SCB-27C and SCB-125 schemes which included the fitting
of an angled deck, a deck edge lift on the starboard side, a fully enclosed bow and a redesigned
island superstructure. In this guise she was redesignated CVA.38 and remained in service until
1971. (*US National Archives*)

The three large Midway class carriers were all substantially modernised in the 1950s. This is
USS *Franklin D. Roosevelt* (CVA.42) after completion of the SCB-110 programme in 1956.
(*US National Archives*)

The prototype Supermarine 510 flown by Lieutenant J. Elliott RN makes the world's first deck landing by a swept wing aircraft aboard HMS *Illustrious* on 8 November 1950. (*Fleet Air Arm Museum*)

The twin engined Supermarine 508 first flew on 31 August 1951 and is shown here during deck landing trials aboard HMS *Eagle* in May 1952. A swept wing development was the Type 525 which eventually went into production as the Type 541 Scimitar. (*Fleet Air Arm Museum*)

formation of de Havilland Sea Vixen FAW.2s. This was the final production version of the DH.110 ~~se~~ries of twin engined all weather jet fighters, of which the prototype first flew in September 1951. ~~H~~owever production examples did not reach front line squadrons until 1959. (*Fleet Air Arm Museum*)

~~Th~~e supersonic Chance Vought F8U-1 Crusader flew in prototype form in 1955 and entered front ~~li~~ne service less than two years later. It was destined to provide the backbone of the US Navy's ~~fi~~ghter strength until gradually replaced by Phantoms from the mid 1960s onwards. One of the two ~~pr~~ototypes is shown here aboard USS *Forrestal* in 1956. (*US National Archives*)

The McDonnell F3H Demon had a chequered career. The initial Westinghouse powered F3H-1 was a total failure and only when re-engined with a more powerful and reliable Allison J71 turbojet was its potential realised. Thus although the prototype flew in 1951, the re-engined F3H-2 did not reach squadrons until late 1956. One of these is shown being manhandled onto the catapult track aboard the recently commissioned USS *Forrestal* (CVA.59) in 1956. Note the retractable air to air refuelling probe. *(US National Archives)*

The Lockheed XFV-1 was one of two vertical take off fighters designed around the 5,850 shp Allison T40 turboprop. While the Lockheed design never demonstrated the transition from vertical to horizontal flight, the rival Convair XFY-1 successfully achieved vertical take offs and landings and transition to and from horizontal flight. However both projects proceeded no further than the prototype stage. (*US National Archives*)

Another blind alley was the concept of a waterborne jet fighter. The twin engined Convair XF2Y-1 Sea Dart employed a hydro-ski as opposed to a more conventional planing hull but the project was cancelled in 1956. (*US National Archives*)

The Douglas XF4D-1 Skyhawk was revolutionary for its time – offering jet performance with a useful load carrying capacity, all in an aircraft weighing in at half that of other contemporary designs. First flown in June 1954, almost 3,000 Skyhawks were built and some remain in service today. (*US National Archives*)

he US Navy finally achieved a credible nuclear capability with the twin engined Douglas A3D-1
ywarrior jet bomber. First flown in October 1952, the Skywarrior was by far the largest and
aviest naval combat aircraft ever produced up to that time. (*US National Archives*)

production Skywarrior aboard USS *Forrestal* in 1956 demonstrates the folding tailfin as well as
lding wings to enable the aircraft to fit in the standard carrier hangars. (*US National Archives*)

The ultimate British naval jet bomber was the Blackburn Buccaneer which first flew in 1958 although its origins went back to a far sighted Admiralty requirement issued in 1951. *(Fleet Air Arm Museum)*

The Fairey Gannet ASW aircraft was unique in being powered by the Double Mamba turboprop driving contra-rotating propellers. One half of the engine could be shut down to extend range and endurance. The Original Fairey GR.17/Type Q prototype provided accommodation for a pilot and observer but production aircraft featured an additional cockpit for a third crew member. (*Fleet Air Arm Museum*)

The Grumman F9F-7 Cougar was a swept wing development of the F9F-5 Panther and offered a considerable increase in performance. Although the prototype flew in 1951, Cougars entered service too late to see action in the Korean War. *(US National Archives)*

The Chance Vought F7U Cutlass was an impressive and futuristic looking design although its in service record was marred by numerous accidents. (*US National Archives*)

The three Midway class carriers which entered service between 1945 and 1947 were the largest US carriers built up to that time and were well able to cope with the first generation of jet aircraft. Shown here is the name ship of the class, USS *Midway* (CV.41), in 1949. Note the Sikorsky HO3S helicopter on the foredeck. (*US National Archives*)

a tendency for the airflow to spread spanwise and cause difficulties with lateral control. Taken together, these changes substantially improved handling to the extent that most pilots found the Cougar easier to handle in the carrier environment than its straight-winged predecessor. It should be noted that the problems that the Grumman team experienced echoed those that had affected the British Supermarine Type 510 and Hawker P.1052, and the solutions adopted were much the same.

The Cougar was powered by a 7,000 lb thrust Pratt & Whitney J48 turbojet, which had also been installed in the later Panther variants and was actually a development of the Rolls-Royce Nene of which the British equivalent was named the Tay. However, by the early 1950s new British jet fighters were being designed around axial flow engines such as the Avon and Sapphire and the only British application of the Rolls-Royce Tay was in an experimental jet-powered version of the Viscount airliner (Type 633), which flew in March 1950 but did not enter production. On the other hand the Cougar proved extremely adaptable and ultimately a total of 1,988 were built, the last being delivered as late as February 1960. Although too late to see service in the Korean War, the Cougar rapidly replaced Panthers and Banshees in some twenty US Navy squadrons, these all receiving the F9F-6 and -7 versions. In a parallel with Panther experience, the F9F-7 was powered by an American-designed Allison J33-A-16 turbojet rated at 6,350 lb thrust. However, most of these were eventually refitted with J48s and the last fifty produced were completed with the more powerful Pratt & Whitney engine.

Once the basic F9F-6 was established in production, Grumman had time to look at improving the design and this resulted in the F9F-8, which first flew on 18 January 1954. This incorporated several changes to the wing, including a thinner profile, increased wing area, and extended and cambered leading edges. Additional fuel tankage was incorporated in the extended leading edges and the fuselage tank was enlarged, increasing total capacity from 919 to 1,063 US gallons. The wing modifications resulted in a useful increase in critical Mach number and the additional fuel increased range by almost 300 miles. However, the extra weight reduced the rate of climb and service ceiling but this was offset to some extent by much improved handling and manoeuvrability. There were several sub variants of the F9F-8, including a photo reconnaissance version (F9F-8P), which flew in February 1955, and a two-seat trainer (F9F-8T), which flew in February 1956. Finally, some Cougars were modified for the tactical nuclear strike role with the fitting of LABS as in the FJ-4B Fury and these were designated F9F-9B. Although fighter versions of the Cougar were phased out of frontline service with the Atlantic and Pacific Fleets service by 1959 in favour of more advanced types, the F9F-8P was operational until 1960. The trainer version (later redesignated TF-9J) served with five training squadrons and the last of these, VT-4, did not relinquish its Cougars until 1974. In addition, many reserve units continued to fly Cougars throughout the 1960s.

Although too late to see combat, the swept wing Grumman Cougar and North American Fury provided the backbone of the US Navy's carrier air groups in the years following the Korean War. As already related, both were developed from straight-wing, first-generation jets under the impetus of the challenge presented by the Russian-designed MiG-15. However, by the early 1950s the US Navy had other projects underway for even more advanced aircraft, although none of these would see full-scale service until the latter half of the decade. The benefits of swept wings had become apparent when allied engineers and scientists gained access to the results of German wartime experience and research, but there were other advanced aerodynamic features that the Germans had applied and a number of US designers attempted to make use of these. Foremost amongst these was the concept of the delta wing pioneered by Dr Alexander Lippisch and when the US Navy needed a fast-climbing interceptor, this configuration appeared to offer some significant advantages. During World War II, the principle of a standing Combat Air Patrol (CAP) was established with waiting interceptor fighters being directed by radar onto incoming targets. However, the new jet fighters had limited endurance so a standing airborne CAP was difficult to organise and costly in fuel consumption. Perhaps more significantly, the performance gap between fighters and jet bombers had narrowed considerably to the extent where the bomber would be difficult to catch in a pursuit scenario. From this problem evolved the concept of a Deck Launched Interceptor (DLI), which remained on short-notice standby on the carrier's deck. Assuming an inbound missile-armed bomber flying at 550 mph at 40,000 feet was detected by radar at a range of 100 miles, a further nine or ten minutes would elapse before it was close enough to launch its missiles. If the DLI was at five minutes' notice, this left less than four minutes for it to launch, climb to 40,000 feet and destroy its target. This implied a minimum rate of climb of around 15,000 feet/min, as well as high speed and good manoeuvrability at that altitude.

To meet this requirement the Douglas company proposed a delta-wing fighter powered by a Westinghouse J40 axial flow turbojet, which was expected to provide 7,000 lb thrust, increasing to 11,600 lb with afterburning. A development contract was awarded in June 1947 but as design work got underway the wing planform was progressively modified. Instead of a pure delta wing, the final configuration was more akin to a tailless aircraft with low aspect ratio wings having a sharply swept leading edge and rounded wingtips. The pilot sat well forward with bifurcated leading edge intakes just behind the cockpit. A tricycle undercarriage with a long nosewheel oleo for catapult launches was fitted, and the resulting high angle of attack on the ground required a small retractable tailwheel to guard against excessive pitch up at launch. The standard armament was four 20 mm cannon – a missile armament not being contemplated at this time. With the design finalised, a contract for the construction of two prototype XF4D-1 Skyrays was awarded in December 1948, although the first of these did not fly until 23 January 1951. Even then flight testing was limited due to the fact that the intended Westinghouse J40 engines were not ready and both prototypes were initially powered by Allison J35-A-17 turbojets, which were only rated at 5,000 lb thrust so that the full performance envelope could not be demonstrated.

As will be seen, the Skyray was not the only aircraft to suffer from problems with the J40 engine, which also formed the basis of several other contemporary projects. Even when production examples of the engine became available, they proved to be very unreliable with an alarming tendency to shed turbine blades in flight. Eventually the whole engine project was cancelled in 1953. Nevertheless, the availability of early J40s allowed the Skyray to demonstrate its potential, which it did in spectacular fashion on 3 October 1953 by wresting the absolute world airspeed record from Britain (set by a Supermarine Swift on 26 September 1953 with a speed of 735.7 mph). The average speed achieved by the Skyray over a 3 km course was 752.944 mph. It is interesting to note that both records were set at very low level in very high ambient temperatures where the speed of sound would be in excess of 760 mph so that Mach numbers in the region of 0.98 were attained. In fact, neither aircraft was supersonic in level flight at that time, although they could exceed Mach 1.0 in a dive, especially in colder air at high altitude where the speed of sound reduced to around 660 mph. Despite this success, the cancellation of the J40 engine programme forced the Douglas team to find a suitable alternative and this was to be the Pratt & Whitney J57-P-2. In fact, this was a much better engine, offering 9,700 lb thrust, which could be boosted to 14,800 with afterburning, and was to prove a very reliable powerplant. The later J57-P-8 offered 10,200 lb dry thrust and 16,000 lb with afterburning. Although the physical integration of the new engine with the Skyray airframe posed few difficulties, trials with the first production aircraft (first flight 5 June 1954) revealed aerodynamic problems with the air intakes, which were eventually solved by the addition of a splitter plate between the fuselage and intake, a device applied to many subsequent supersonic aircraft. However, this added further delays to the Skyray's service debut and it was not until mid 1956 that it finally became fully operational, five years after the first prototype and almost two years after the first re-engined production aircraft had flown (although initial carrier trials were carried out by one of the J40-powered prototypes aboard the USS *Coral Sea* in October 1953). Nevertheless, the US Navy now possessed a remarkable interceptor. With the J57 engine, the Skyray was now just supersonic in level flight at altitude and had an initial rate of climb in excess of 18,000 feet/min. In May 1958 a Skyray set a series of time to height records, reaching 15,000 metres (49,212 feet) in only two minutes and thirty-six seconds! The capturing of the world airspeed record was the first time that this had been achieved by a carrier-capable aircraft and the production F4D-1 was the US Navy's first supersonic fighter, although only just so. Outside the time scale of this book, it is interesting to note that Douglas produced an advanced version of the Skyray, which was the F5D-1 Skylancer. The prototype flew in 1956 and reached speeds of Mach 1.5 at 40,000 feet, as well as possessing a limited all-weather capability, but was not ordered into production.

The US Navy could not be accused of being unwilling to try out new concepts. Even before the Skyray development was initiated, it had commissioned the construction of three prototype fighters from Chance Vought in June 1946. These were again based on German work and the result was a revolutionary tailless design with directional control achieved by twin fin and rudder assemblies at mid span on each wing. Like the Skyray, the new XF7U-1 Cutlass was intended as a deck-launched interceptor and to achieve the required performance a twin-engine configuration was adopted, two Westinghouse J34-WE-32 turbojets with afterburning being set side by side in the fuselage nacelle. The result was unlike anything flown before or since and, considering the unorthodox layout, performed better than might have been expected, although several serious problems were encountered. The prototype XF7U-1 flew on 29 September 1948 and was followed by the first of fourteen production F7U-1s in March 1950. These were eventually allocated to the Navy's Advanced Training Command at Corpus Christi in 1952 as problems with the J34 engine as well as handling difficulties made the Cutlass unsuitable for carrier deployment. These caused the cancellation of the F7U-2 with more powerful J34s and development was centred on the F7U-3, which was re-engined with Westinghouse 4,660 lb thrust J46-WE-8A engines. The centre fuselage was entirely re-designed and the underwing section of the vertical tail surfaces was virtually eliminated. The F7U-3 was first flown in December 1951 and initially four Navy squadrons (VF-81, VF-83, VF-122, VF-124) were equipped with the Cutlass, although operational deployments did not commence until May 1954, the first being VF-81 aboard the USS *Ticonderoga*. A total of 290 F7U-3s were delivered, which included ninety-eight F7U-3Ms with provision to carry four Sparrow air-to-air missiles and twelve F7U-3P photo reconnaissance aircraft. However, the type's record in service was very poor with serious maintenance problems and a frighteningly high accident rate so that production ceased in 1955 and it had been entirely replaced before the end of the decade. Originally intended as a pure interceptor, it was out-classed in this role by the Skyray and subsequently was more often utilised in the attack role, carrying two 1,000- or 2,000 lb bombs, and it could also be fitted with a centreline pod carrying forty-four 2.75 inch air-to-air unguided rockets.

Although the US Navy was never able to field a swept wing jet fighter in the Korean War, things might have been different if development of another design had progressed as hoped. This was the McDonnell F3H Demon, which was ordered in prototype form in September 1949 following the company's response to an earlier request for proposals issued by the Bureau of Aeronautics in May 1948 for a swept-wing naval inter-ceptor. For its time, the Demon was an advanced design with sharply swept flying surfaces, wide lateral air intakes and pro-vision for afterburning with the tailplane and fin being set well above the jet efflux. The prototype XF3H-1 flew on 7 August 1951 but even by that time the programme was in deep trouble. Things were not helped by a BuAer requirement that the Demon should be redesigned as an all-weather fighter under the designation F3H-1N. This significantly delayed develop-ment, the first production examples not flying until December 1953. In the meantime a much more serious problem had arisen in the shape of the failure of the Westinghouse J40 engine programme. The afterburning 9,200 lb thrust J40-WE-8 had initially been selected as the powerplant for the Demon but the XF3H-1 prototype was only fitted with the unreheated 6,500 lb thrust J40-WE-6. Eventually the afterburning variants were fitted to both prototypes and initial carrier trials were carried out aboard the USS *Coral Sea* in October 1953. Even by that time the J40 was proving unreliable, resulting in damage to one aircraft, and both being grounded for various periods. When testing of production F3H-1Ns began, the results were even more disastrous with no fewer than five aircraft being destroyed in accidents, including three in which the pilot was killed. In most cases the root cause of the accident was a failure of the J40 engine and it was quite clear that the Demon was seriously underpowered. Matters were so bad that many of the fifty-eight F3H-1Ns completed were never even flown, but were shipped by barge down the Mississippi from St Louis to Memphis where they were used as instructional airframes.

The US Navy had a major commitment to the J40 engine, which was the prime powerplant for several projected aircraft (including the F4D Skyray and twin-engined A3D bomber), and was reluctant to abandon development despite the evidence of flight tests. However, the McDonnell team was determined to salvage the Demon programme and persuaded the US Navy to allow two F3H-1N airframes to be converted to accommodate an Allison J71-A-2E rated at 9,700 lb thrust (14,400 lb with afterburning). Thus powered, the first F3H-2N flew on 23 April 1955 and proved to be a much better performer, although the Demon would have benefited further if an even more powerful engine could have been fitted but this was not possible without a major redesign. Eventually the F3H-2N began to reach squadrons in late 1956, subsequently embarking in the USS *Forrestal* for a Mediterranean deployment in January 1957. The standard F3H-2N was armed with fuselage-mounted 20 mm cannon but in August 1955 the missile-armed F3H-2M made its first flight. This could carry four AAM-N-2 Sparrow 1 semi-active homing guided missiles, which relied on the target being illuminated by the aircraft's AN/APG-51B radar. In this guise the Demon became the US Navy's first all-weather missile armed interceptor and remained in service until replaced by more capable fighters in the early 1960s. Thus the Demon was too late to see service in the Korean War and had been phased out of service by the time the United States became involved in Vietnam.

Despite the fact that the Demon had a relatively inauspicious career, it did provide the springboard for another project that was to become not just a first-class naval fighter, but one of the most successful all-round combat aircraft ever flown – the Mach 2 capable McDonnell F4H Phantom II. Although the first flight and subsequent development of this superb aircraft lie outside the period covered by this book, its initial development can actually be traced back as far as 1953 when McDonnell began a series of studies aimed at offering a substantial improvement in capability and performance of the F3H Demon. Under the designation Model 98, a series of single- and two-seat, single and twin-engined proposals were made. Of these, and after discussions with the US Navy, the Model 98B powered by either two Wright J67s or two General Electric J79s was taken as the basis for a new single-seat aircraft initially designated AH-1 in recognition of its planned attack role. However, by May 1955 when the construction of two prototypes was authorised, it was decided that they would be completed as two-seat, missile armed, all-weather fighters under the designation YF4H-1. Subsequently the prototype flew on 27 May 1958 and its startling performance led to substantial US Navy orders and, in a remarkable achievement, it was also ordered in quantity for the USAF – the first time that a naval fighter had achieved this distinction.

In fact, successful as the Phantom was to prove, it would not be the US Navy's first fully supersonic fighter as it was preceded by two other types, both of which were easily capable of exceeding Mach 1 in level flight. The first of these to enter service was the Grumman F11F-1 Tiger, which had first flown on 30 July 1954 and subsequently entered operational service in early 1957. Development of the F11F had commenced in December 1952 when the possibility of applying the area rule principle to the swept wing F9F was investigated. It quickly became apparent that a fresh design would be required and this evolved as a slim-fuselaged, swept wing jet fighter with an empty weight of around 13,000 lb. The incorporation of area rule led to distinctive narrowing of the fuselage where the wings were mounted in order to eliminate rapid changes in the total cross-sectional area presented to the airflow. The slim fuselage was achieved by using an axial flow jet engine, in this case an afterburning Wright J65, which was actually a licence-built version of the British Armstrong Siddeley Sapphire. Consider-able attempts were made to reduce weight and one unusual feature was that the wingtips were folded down manually, saving the complexity of the more conventional power-actuated upward-folding wings. The first prototype was lost in an accident in October 1954 due to an engine flameout – a recurrent problem with the J65. However, the second prototype had flown by then and was able to carry on the test programme, exceeding Mach 1 for the first time before the end of the year. A third prototype flew in December 1954 and incorporated various modifications, including redesigned tail surfaces, a longer nose, air intake splitter plates, a new clear-view canopy and had provision for an air-to-air refuelling probe to be fitted. A standard armament of four 20 mm cannon was fitted and the F11F-1 could carry four Sidewinder missiles or two missiles and two drop tanks of 150 US gallons.

Carrier trials aboard the USS *Forrestal* in April 1956 highlighted the need for further modifications, notably increasing the fuel capacity to make good shortfalls in range and endurance. As already related, service entry followed in 1957 but the Tiger's operational career was relatively short and it was withdrawn from frontline units by 1961, although it enjoyed a longer career with second line training units and was adopted as the mount for the famous Blue Angels naval demonstration team. When the Tiger passed out of frontline service, it was the end of an era for Grumman who had provided naval fighters continuously from 1933 and it was not until the 1970s that another Grumman jet fighter (the F14 Tomcat) was to serve aboard US carriers. The reasons for the Tiger's premature withdrawal were varied but included poor handling, which made it an unsteady gun platform, and the unreliability of the J65 engine. However, the main reason was that it was totally eclipsed by another fighter that had entered service at the same time.

This was the Chance Vought F8U Crusader, which originated from a US Navy requirement for a supersonic air superiority fighter issued in 1952. This company had, of course, made its name as the builder of the famous Corsair piston-engined fighter but had experienced less success with its subsequent jet designs. Awarded a contract in May 1953 for further development of its proposals, Chance Vought produced a fairly conventional design based around a single Pratt & Whitney J57-P-11 turbojet, which could produce 18,900 lb thrust with afterburning and was positioned towards the rear of the long fuselage. A pointed radome with a semi-circular chin air intake below gave the Crusader a distinct profile but its most unusual feature was the high-mounted swept wing with noticeable anhedral. This was unusual in a jet fighter but was accounted for by a unique feature intended to make the task of landing this hot ship aboard a carrier easier than it might otherwise have been. In normal flight the top surface of the wing centre section lay flush with the top of the fuselage but at slower speeds the leading edge could be raised by means of screw jacks, so increasing the angle of incidence of the whole wing relative to the fuselage. This in turn effectively reduced the nose-up attitude of the whole aircraft, improving the pilot's view of the deck while also increasing lift. The high wing layout did mean that the main undercarriage legs had to retract in the fuselage, which resulted in a relatively narrow wheel track.

In service the F8U Crusader offered a substantial increase in performance over previous naval fighters. The top speed was around Mach 1.7 at altitude, although the initial rate of climb was only 12,000 feet/min, still very good but well below that achievable with the Skyray. This difference was accounted for by the weight of the Crusader, which grossed at 34,000 lb compared with the 25,000 lb MAUW of the Skyray. However, the Crusader had a much better endurance and was capable of carrying a wide variety of weapons apart from the internal four 20 mm cannon. The standard fit was either two (later four) Sidewinder infra-red homing homing missiles, or up to 5,000 lb of bombs or rockets, or any combination of these. Given the advanced performance of the Crusader, its development and entry into service was completed in an amazingly short time, especially when compared with earlier jets such as the Skyray and Demon. The prototype XF8U-1 flew in March 1955 and production standard aircraft were coming off the assembly lines by the end of 1956 with the first squadron (VF32) forming in March 1957. By the end of the year the squadron was deployed aboard the USS *Saratoga* and subsequently various versions of the Crusader flew with no fewer than seventy US Navy and USMC squadrons in a career that lasted until the mid 1980s. A total of 1,261 Crusaders were produced, which included forty-two for the French Navy, one of the few cases where US Navy fighters were exported. One of the most important variants was the RF-8 in which a battery of five cameras replaced the 20 mm cannon and provision was made for additional fuel. Both fighter photo reconnaissance versions saw considerable action during the Vietnam War, some units serving right through to the end in 1973.

The variable incidence wing of the Crusader was one example of altering wing characteristics to overcome the problem of landing high-performance swept wing aircraft on a carrier deck. A more ambitious solution dated as far back as 1946 when Grumman were asked to build a research aircraft to test swept wings with an eye to their later application to the XF9F-2 Panther. The initial proposal was for a single-engined, high-mounted, modified delta-wing aircraft with a T-tail and a contract was issued in April 1948 for two prototypes. However, by 1950 a new specification was issued calling for a fighter that would have good low-speed handling qualities and be capable of transonic speeds. At the same time a heavy armament and long range was required so that the final Grumman design increased in weight from around 18,000 lb MTOW to over 31,000 lb. After investigating the possibility of using variable incidence, the Grumman team went a step further and proposed a variable sweep wing with full-span leading edge slats and flaps along 80 per cent of the trailing edge.

The prototype XF10F-1 Jaguar eventually flew on 19 May 1952 but almost immediately severe problems were manifest. These were both technical failures and stability problems, although the wing sweep mechanism itself caused no trouble and turned out to be very reliable. The Achilles' heel of the project was the use of the ill-fated Westinghouse J40 engine. The cancellation of this engine in 1953 and the subsequent grounding of all aircraft fitted with it effectively ended the Jaguar programme. Nevertheless it should be recognised as a pioneering attempt at what was actually a viable proposition and the concept of variable sweep was incorporated in the American F-111 and B-1 bombers, and the European Tornado.

At this point brief mention should be made of two significant aircraft that were built in prototype form to explore the concept of a VTOL combat aircraft. These were the Convair XFY-1 and the Lockheed XFV-1. Both came about as a result of studies made by both the US Navy and Air Force in the late 1940s and with the outbreak of the Korean War development funds became available. The two designs were very similar in concept, power for take-off and conventional flight being provided by turboprops driving a pair of three-bladed contra-rotating propellers. With broad chord blades these acted as helicopter rotors to lift the aircraft from the ground, the transition to horizontal flight being made gradually as speed and altitude increased. The Lockheed XFV-1 had short stubby wings with wingtip tanks, and a cruciform tail assembly that incorporated four wheels on which the aircraft rested when on the ground. The pilot was provided with an ejector seat. For initial flight trials a 5,850 hp Allison XT40-A-6 turboprop was installed with a more powerful YT40-A-14 scheduled for later tests, although this never actually became available. The production FV-2 would have had an even more powerful T54-A-16. The proto- type was fitted with a temporary fixed undercarriage to permit conventional take-off and landings for the initial test flights and officially flew on 16 June 1954, although it had previously become airborne in high-speed taxi trials. Although transitions from horizontal to vertical flight modes and back again were made in the air, vertical take-offs and landings from the ground were never attempted as the trials engine did not produce enough power. The programme was cancelled in mid 1955 when it was realised that even if full transitions could be achieved, the performance of the turboprop fighter would lag well behind that of contemporary jets.

The Convair XFY-1 was much more successful and featured a broad delta wing, together with upper and lower vertical fins. This made a much more stable base (an important con- sideration aboard ship) but in other respects the design was similar to that of the Lockheed aircraft with the same power- plant and a tilting ejector seat for the pilot. Initial tethered flights commenced in April 1954 and these culminated in the first free flight in which transition to horizontal mode was successfully accomplished and was then followed by a vertical landing. This occurred in November 1954 and subsequently the XFY-1 recorded a maximum speed of 610 mph at 15,000 feet and had an excellent rate of climb, reaching 30,000 feet in 4.6 minutes. Despite the apparent success of the programme the project was cancelled in 1956, one possible reason being that the widespread adoption of VTOL combat aircraft might result in a reduction in the size of the US Navy's carrier fleet – something that many Admirals did not wish to see. A similar attitude prevailed in Britain in the 1960s when aircraft such as the P1127 and Kestrel (forerunners of the Harrier) were seen by some as a possible threat to the Royal Navy's carriers.

Convair were to achieve considerable success with the production of delta-winged jet fighters for the USAF (F-102 and F-106) and applied some of their experience to produce a jet-powered water borne fighter, the XF2Y-1 Sea Dart. Powered by two 3,400 lb thrust J34-WE-32 engines, the Sea Dart was very similar in outline to the F-102 Delta Dagger except that the engine intakes were on the top of the fuselage to prevent spray ingestion on take-off. Instead of a conventional flying boat hull, the XF2Y-1 was fitted with retractable hydro skis, which offered less drag in the air and permitted higher speeds on the water. The official first flight was on 9 April 1953, although the aircraft had previously been briefly airborne during taxi trials. There were problems both with the engines, which failed to develop the expected thrust, and with controlling the aircraft on take-off while running on the skis. Consequently only three development YF2Y-1s were built and one of these crashed in November 1954. Although the US Navy continued test and evaluation trials, the project was finally halted in 1956.

By the mid 1950s, the development of jet combat aircraft for naval use had advanced considerably in the decade following the end of World War II. By 1955 high-performance swept wing aircraft, some with supersonic capability, were either in service or under active development in both Britain and America. However, it was almost inevitable, given the strength of their industrial base, that the Americans should forge ahead. Thus in 1955, while the Royal Navy was still operating straight-winged Seahawks and Sea Venoms, the US Navy had swept wing Furies and Cougars and was about to introduce the fast-climbing Skyray, while the supersonic Crusader (capable of speeds in excess of 1,000 mph) had already flown in prototype form.

7

ATTACK, REPEAT, ATTACK

The title of this chapter is the text of the message sent by Admiral Halsey to Admiral Kinkaid, OTC carriers *Enterprise* and *Hornet*, after hearing that scout aircraft had located the Japanese carrier force in the opening moves of the Battle of Santa Cruz in late October 1942. In the decade after the end of World War II this could well have been adopted as the US Navy's motto as it sought a principal role in the new era of atomic weapons. During most of World War II the carrier's main function had been to provide strike forces to sink enemy ships while also defending friendly forces against enemy air attacks. As the Pacific War reached its closing stages and in the absence of any significant naval opposition, the emphasis began to change to the ability to attack land-based targets and support friendly forces ashore. In the post-war era the US Navy had no immediate maritime rivals and so the land attack mission, or power projection, became even more important. This was especially so with the coming of the atomic bomb, which threatened the very existence of naval task forces as well as appearing to establish the strategic bomber flown by the Air Force as America's principal weapon.

The problems and tactics associated with attacking land targets were diverse in nature and required different solutions. The tactical role of supporting troops ashore required specialised aircraft whose main attributes were the ability to carry heavy loads and a wide variety of ordnance. As carrier fighters would provide an escort and could assist in suppressing anti-aircraft fire, out and out performance was not an issue. In the immediate post-war era this task was carried out by aircraft such as the Corsair and Helldiver and these were supplemented, and eventually replaced, by the purpose-designed Skyraider and

Mauler. Eventually the AD Skyraider became the most prolific and, as related in Chapter 3, had an active career well into the Vietnam War era. However, even as the Skyraider was entering production in June 1945, steps were already being taken to secure a successor. While at that time the first jet fighters were being produced, it was thought that a turboprop would be a more suitable powerplant for an attack aircraft and Douglas drew up several designs based on the Skyraider but initially none of these went past the project stage. However, in 1947 Douglas was instructed to go ahead with two prototypes that would act as test beds for the turboprop concept and have the potential to be developed as an attack aircraft. The new design was designated XA2D-1 and was subsequently given the name Skyshark. The original proposal was to modify the basic Skyraider airframe to take an Allison XT40-A-2 turbine, driving nose-mounted three-bladed contra-rotating propellers (the same combination was used in the Lockheed and Convair VTOL projects described in the last chapter). The Allison XT40 consisted of two Allison T38 turbines driving the propellers through a common gearbox and the maximum power output was 5,100 eshp. In practice it was found that substantial changes were necessary to accommodate the turboprop and to cope with the much higher installed power so that the XA2D-1 ended up as a completely new aeroplane, bearing only a passing resemblance to its predecessor. The name Skyshark was quite apposite as the huge spinner for the propellers, coupled with the twin air intakes mounted low down on either side of the nose gave the aircraft a very predatory appearance.

The prototype XA2D-1 flew on 26 May 1950 and substantial improvements over the piston-engined Skyraider were quickly demonstrated. It was much faster with a maximum speed of just over 500 mph, the rate of climb was more than doubled and the range was increased by around 30 per cent to a maximum of 2,200 miles. Carrying a similar bomb load to the Skyraider, its efficiency in terms of delivering a given weight of ordnance over a specified distance was up by over 50 per cent. Unfortunately these advantages came at a cost as the complex power train proved unreliable and led to two accidents in which one pilot was killed, although the other successfully ejected. This last accident occurred in 1954 and shortly afterwards the orders for 339 Skysharks were cancelled and only the two prototypes and six production aircraft were completed and flown.

Although problems with the Allison turboprop and the transmission train were a significant factor in the cancellation of the Skyshark programme, the main reason was that something much better was now on the horizon. This was another Douglas project, the jet-powered A4D-1 Skyhawk, which was destined to become one of the most successful combat aircraft ever built. In fact, it remains in service in small numbers even today. Like many successful aircraft, it began life as a private venture project rather than as a result of any official specification. Ed Heinemann, Douglas' chief designer, had realised that jet fighters were increasing in size and weight almost uncontrollably and he set out to reverse this trend. He put his ideas to the US Navy early in 1952 but at that time the service already had several other fighters under development (such as the Skyray and Demon) but thought that similar design principles might be applied to an attack bomber. The US Navy requirement called for an aircraft with a speed of 500 mph and a combat radius of 345 miles with a 2,000 lb bomb load. The maximum gross weight should not exceed 30,000 lb. Within a very short time scale Heinemann produced a sketch design for a small delta-wing single-engined jet bomber with a top speed of 600 mph and a combat radius of 450 miles. The most amazing aspect of the design was not that it handsomely exceeded the US Navy's requirements but that it did so at a normal loaded weight of only half the expected 30,000 lb. Not for nothing was the Skyhawk nicknamed 'Heinemann's Hot Rod'! Despite misgivings, the US Navy gave the go-ahead for further work and a prototype was ordered on 21 June 1952.

To save weight and eliminate the need for a complicated wing-folding mechanism a short span delta wing was adopted and this was complemented by a delta tailplane set low down on the swept tail fin. The chosen powerplant was the Wright J65 (a licence-built Armstrong Siddeley Sapphire) and despite problems experienced with this engine in other aircraft (notably the F11F-1 Tiger) it gave no trouble in the A4D and over 1,300 J65-powered Skyhawks were delivered. Eventually it was replaced by the slightly more powerful but more efficient Pratt & Whitney J57 but this did not occur until the A4D-5 (redesignated A-4E in 1962) flew in 1961. Prior to that the proto-type XA4D-1 had flown on 22 June 1954, only one day over two years from the time it was ordered. The flight test pro-gramme proceeded equally rapidly and revealed only minor problems, which were easily rectified. Carrier qualification trials took place aboard the USS *Ticonderoga* in September 1955 and the first operational squadrons began receiving aircraft some twelve months later. Ultimately some 2,960 Skyhawks were built and in addition to serving with over 100 US Navy and USMC squadrons, it also served aboard Argentine, Brazilian and Australian carriers and was sold to the Air Forces of New Zealand, Israel, Singapore, Malaysia, Indonesia and Kuwait.

When originally ordered, the US Navy specified that the Skyhawk should be capable of carrying nuclear weapons and this illustrates a vital component of US Navy policy in the post-war era and one that had a substantial influence on carrier design and the characteristics of many new-generation air-craft. The dropping of the first atomic bomb on Hiroshima on 15 August 1945 completely changed the nature of warfare for ever and proponents of the strategic bomber were quick to take advantage of the fact to support their case. When the USAF officially came into being as a separate force (having previously been part of the US Army) in 1947 it was naturally intent on consolidating its position by being the sole custodian of the nation's nuclear strike capability. In addition, many argued that the devastating power of a single atomic bomb rendered the traditional carrier task force vulnerable to destruction. Naturally, the naval staff did not accept these arguments and were determined that their service would continue to have a major role in American defence policy, including the delivery of nuclear weapons to strategic targets, arguing that the proven flexibility and mobility of aircraft carriers would be a vital asset even in an atomic war.

During World War II, naval aircraft typically carried nothing larger than a 2,000 lb bomb as a weapon of this size was more than capable of sinking a ship and was useful in the tactical support of ground troops. However, for attacking strategic targets such as factories and oil refineries, much larger bombs were required and by 1945 land-based bombers could carry bombs up to 12,500 lb or even larger. If a carrier-based bomber was to have any credibility in the strategic role, then it would have to carry similar-sized weapons and it was no coincidence that this was also the size of a typical first-generation atomic bomb. The US Navy proposed a gradual approach to building up a suitable carrier-based bomber force by specifying a succession of aircraft with steadily increasing performance and sophistication. The first step would be a piston-engined bomber with an empty weight of around 30,000 lb and a speed of over 360 mph at sea level. The radius of action would vary according to the bomb load but with an 8,000 lb bomb would be around 300 nm. Next would come a turboprop-powered 45,000 lb bomber with a combat radius of 500 nm and a speed of 500 mph at altitude. Finally, there would be a jet bomber with a radius of 2,000 nm with a 12,000 lb bomb, this being sufficient to ensure that any target within the Soviet Union could be reached by a carrier-based bomber. The stumbling block with this project was that it would be a very large aircraft with a take-off weight in excess of 100,000 lb. While the first two would be able to operate from suitably modified Essex and Midway class carriers, an entirely new and larger carrier would be required for the largest bomber and it was this parameter that drove forward the case for the USS *United States* (CV58 – see Chapter 6).

The initial requirement was met by the North American AJ-1 Savage of which three prototypes were ordered as early as June 1946, an indication of the priority that the US Navy accorded to the nuclear strike role. Initially this was to be powered by two 2,400 hp Pratt & Whitney R-2800 radial piston engines but at an early stage it was decided to install an Allison J33 turbojet in the rear fuselage. Rated at 4,600 lb thrust, the jet was intended to provide a high-speed dash capability over the target, giving it a maximum speed of 471 mph, and to boost take-off performance. With a single 12,000 lb Mk.4 nuclear bomb, and fuel for a combat radius of just under 1,000 nm, the Savage weighed in at over 50,000 lb, making it by far the heaviest carrier aircraft ever operated up to that time and the jet boost was certainly welcome. The first XAJ-1 flew on 2 July 1948 and subsequently fifty-five production-standard AJ-1s were delivered, commencing in May 1949. The flight test programme was not without mishaps and two of the three prototypes were lost in accidents but in September 1949 the first operational squadron, VC-5, was formed and carrier trials aboard the USS *Coral Sea* followed in April 1950.

The next version of the Savage was the AJ-2P. This was a specialised photo reconnaissance version fitted with no fewer than eighteen cameras, which could be used for day or night photography at high and low altitudes. Photo flash bombs were carried in the reduced-capacity weapons bay and a modified nose carried the electronic equipment necessary to control the battery of cameras. Additional fuel was carried to increase range and external changes included a taller tail fin and no dihedral on the tailplane, these latter modifications being intended to overcome handling problems highlighted in the test programme. The AJ-2P first flew in March 1952 and subsequently equipped two squadrons (VAP-61 and VAP-62). The bomber version was designated AJ-2. This variant first flew in February 1953 and was to serve with a total of ten units at various times. In fact, the Savage's career as a nuclear strike bomber was relatively short and it was phased out of this role from 1956 onwards. Even before then, however, its size and payload characteristics made it suitable for conversion to the air tanker role, this being done by fitting extra fuel tanks and installing a hose and reel unit in the bomb bay. As an aside it is interesting to note that once again the US Navy showed a preference for British inventions, this system of air-to-air refuelling using a drogue at the end of trailing hose having been perfected by Flight Refuelling Ltd. This was in contrast to the USAF, which adopted the retractable trailing rigid boom method.

The AJ-1 and -2 Savage was rushed into service to provide the US Navy with its nuclear capability. However, throughout its service career it was plagued with problems and was not a popular aircraft to fly, not least because the pilots were well aware that if they missed the wires, the crash barriers could not hold the heavy bomber and prevent it ploughing into the deck park ahead. The tightly cowled engines were prone to overheating when climbing at full power and fire warnings were a constant hazard. The Savage was always intended as an interim nuclear bomber and a more powerful version powered by two turboprops and designated XA2J-1 was flown but offered little improvement and this project was cancelled in June 1953. As it was, the performance of the AJ Savage was not good enough to evade interception by jet fighters, especially something as efficient as the MiG-15, but fortunately something much better was on the way and this also explains the early demise of the XA2J.

The new aircraft was the Douglas A3D Skywarrior swept wing twin-jet bomber, which at last provided the US Navy with a high-performance bomber, giving it a true carrier-based strategic nuclear capability. The original Navy Request for Proposals dated back to 1947 and specified an aircraft capable of carrying a 10,000 lb nuclear weapon and able to make a return flight to a target located 2,000 nm miles away. The US Navy did not anticipate that this could be achieved by an aircraft weighing in at less than 100,000 lb and it was this parameter that was the driving force behind the new super carrier project (CV58, USS *United States* – see Chapter 6). However, Ed Heinemann and the Douglas team came up with a proposal for a 68,000 lb aircraft that would be capable of operating from the existing Midway class carriers. This turned out to be fortunate for the US Navy, which lost the political battle with the Air Force, and consequently, as already related, the super carrier was cancelled. Work on the Douglas A3D continued and a contract for two XA3D-1s was awarded in March 1949. These were to be powered by two of the ill-fated Westinghouse J40s rated at 7,500 lb thrust and these were fitted to the prototype, which flew on 28 October 1952. The failure of this engine programme has already been mentioned but, in fact, Douglas was able to turn the situation to its advantage. With the original engines the Skywarrior would have been underpowered but the then revolutionary concept of fitting the engines in self-contained pods attached to underwing pylons meant that it was relatively easy to substitute more powerful 9,700 lb thrust Pratt & Whitney J57 twin-spool turbojets, which also offered improved fuel efficiency. Because the new engines were heavier, the pylons were extended forward to preserve the C of G range and this change also eliminated some aerodynamic problems experienced with the prototype.

Two other factors worked in favour of the new Douglas bomber, one of which was the continued reduction in size of nuclear weapons. When the Savage had been proposed the smallest nuclear weapon, the Mk.4, weighed in at 12,000 lb and when work started on the Skywarrior this had already reduced to 10,000 lb. By the time the latter entered service the Mk.5 bomb with a 120-kiloton yield weighed only 3,300 lb. This meant that either the number of weapons carried could be increased or range could be considerably extended, particularly with in-flight refuelling. The second factor was the introduction of the angled deck, which meant that the new bomber would even be able to operate off converted Essex class carriers if required. However, in practice the existence of the Skywarrior was one of the driving forces behind the new Forrestal class super carriers, which also entered service in 1956.

The Douglas Skywarrior was a very clean design with a high-mounted swept wing with the two turbojets mounted on underwing pylons, which, as well as facilitating the engine change outlined above, also gave much-improved access for maintenance crews. A flight crew of three sat in the forward flight deck, although as a weight-saving measure they were not provided with ejector seats. A novel feature was the provision of a defensive armament in the shape of twin 20 mm cannon installed in a radar-directed remotely controlled turret in the extreme rear of the fuselage. The outer wings folded upwards and the tail fin folded down to allow the aircraft to be stowed in standard-height carrier hangers. The initial production version was the A3D-1 of which forty-nine were built and the first of these entered service with Heavy Attack Squadron One (VAH-1) in March 1956. However, the main production version was the A3D-2 with a strengthened airframe and modifications to permit the carriage of a greater variety of weapons, and these were powered by 10,500 lb thrust J57-P-10s. This version entered service in 1957 but by that time the US Navy had already made a series of long-distance flights with the earlier A3D-1 to advertise the operational capability of the new air-craft. These included a sortie in which aircraft of VAH-1 were launched from the USS *Shangri La* in the Pacific off the Mexican coast and flew direct to NAS Jacksonville on the Florida Atlantic coast without in-flight refuelling.

The subsequent career of the A3D lies outside the scope of this book but it is relevant to note that it proved extremely adaptable with variants being produced for the photo reconnaissance, air refuelling and electronic warfare roles. The A3D was in service throughout the Vietnam War and EW versions remained in service until 1993, having flown missions in the first Gulf War. The basic design also found favour with the USAF, which, as result of experience in the Korean War, raised a requirement for a jet tactical bomber based on the Skywarrior and this eventually evolved into the B-66 Destroyer, which entered service in 1956, the same year as its naval counterpart.

Although the Skywarrior eventually provided the US Navy with a true nuclear bomber, a number of stopgap measures were proposed to supplement the composite-powered AJ Savage until the new jet bomber was available. Inspiration came from the famous Doolittle raid on Tokyo on 18 April 1942. At that point of the war US forces had suffered a series of humiliating defeats in the aftermath of Pearl Harbor and some form of offensive action was urgently required in order to boost the nation's morale. This took the form of a daring operation in which sixteen land-based B-25 Mitchell bombers were hoisted aboard the carrier USS *Hornet* at San Francisco and subsequently flown off when the carrier and her escorting task force were almost 700 miles off the Japanese coast. The bombers achieved complete surprise and, although the material damage inflicted was minimal, the propaganda coup was considerable. After dropping their bombs the B-25s attempted to reach bases in mainland China and almost all were lost in forced landings or when their crews were forced to bale out as fuel was exhausted. The whole operation had been mounted under strict secrecy and when President Roosevelt was subsequently asked where the bombers had come from, he replied 'Shangri La', after a fictitious Chinese haven that featured in a contemporary film. However, in order to commemorate the memory of this famous action the name Shangri La was subsequently allocated to one of the new Essex class carriers.

Echoes of the Doolittle raid can be found in a post-war attempt by the US Navy to provide a credible nuclear strike force. Although the carriage of a 12,000 lb atomic bomb over ranges in excess of 2,000 nm would require the design and development of a new large bomber, the US Navy did, in fact, already possess such an aircraft, although it was never intended for carrier operation. This was the Lockheed P2V Neptune maritime patrol bomber intended for land-based, long-range anti-submarine warfare. The prototype flew in May 1945. One of the first production aircraft was modified by the removal of all armament and having increased fuel capacity in order to establish a new world record for long distance flight. Named 'Truculent Turtle', the Neptune left Perth, Australia, on 29 September 1946. It subsequently landed fifty-five hours and seventeen minutes later at Port Columbus, Ohio, having covered a distance of 11,235.6 miles non-stop without refuelling. This record was to stand for over fifty years.

The Neptune was an exceptionally clean aircraft with an efficient high aspect ratio wing. The original P2V-1 was powered by two 2,300 hp Wright R-3350-8 radial piston engines giving it a top speed of around 310 mph and a maximum range of over 4,000 miles. The P2V-2 had more powerful R-3350-24W engines and one of these aircraft was used to test the feasibility of operating from a carrier in trials conducted aboard the USS *Coral Sea* in April 1948. In order to get this large and heavy aircraft off the deck, rocket-assisted take-off was employed and there was no question of the aircraft landing back aboard. However, as a result of these trials, a total of twelve Neptunes were converted to P2V-3C standard. Powered by two 3,200 hp R-3350-26W engines these aircraft had all defensive armament removed to save weight, carried additional fuel, and were armed with a single 10,000 lb B-4 14-kiloton nuclear weapon. The aircraft were allocated to two squadrons, Composite Five and Six (VC-5, VC-6), and these became operational in September 1948. The Neptunes remained as part of the US Navy's nuclear strike force until 1952 when they were completely replaced with the purpose-designed Savage. The Neptunes could only be operated from the Midway class carriers and while on board they had to be parked on deck where they obstructed any other flight deck movements. Consequently they were only embarked for short periods for training and familiarisation exercises, or at times of heightened tension when US nuclear forces were at a high state of readiness. In general the aircraft were deployed to forward bases from where they could be embarked if required. For example, one squadron was based in Morocco for a while so that the Neptunes were available to a Midway class carrier forming part of the US Sixth Fleet in the Mediterranean. Nevertheless, some impressive demonstration flights were made with the aircraft making flights of over 5,000 miles after a carrier launch to simulate an attack on a strategic target. In the course of one of these flights, P2V-3C made a RATOG assisted take-off from the USS *Coral Sea* on 21 April 1950 at an all-up weight of 74,668 lb, another world record. Although no Neptune ever made a deck landing aboard a carrier, one was experimentally fitted with an arrester hook and carried out dummy deck landing trials at the Naval Test Centre at Patuxent River. Whether or not the Neptune would have had any chance of reaching its targets under war conditions is debatable. As a piston-engined bomber it would have been easy prey to jet fighters and any mission would have needed a considerable element of surprise in order to achieve success. Like the Doolittle raid of 1942, the mission would almost certainly have been a suicidal one-way affair but this is perhaps an indication of how desperate the US Navy was to maintain a nuclear strike capability to justify the existence of the carrier fleet.

As already recorded, the nuclear bomber role was eventually taken over by the jet-powered Skywarrior and, later, by the supersonic North American A-5 Vigilante. Also, throughout the 1950s, the decreasing size of nuclear weapons made it possible for virtually all the US Navy's frontline fighters and attack aircraft to be modified for the strike role and the need for a dedicated bomber diminished. Finally, of course, the strategic nuclear role was gradually taken over by missile-equipped submarines from 1960 onwards.

In the case of the Royal Navy, there was never any serious ambition to take on the strategic nuclear strike role, although by 1951 a formal requirement for the deployment of tactical nuclear weapons was raised. In the meantime the British lacked a suitable dedicated carrier-based bomber, preferring to use multi-role aircraft such as the Sea Fury, FB.Mk.11 and the Firefly FR.4 and the later Attacker and Seahawk, which were both produced as fighter-bombers. Nevertheless, the Admiralty had considered the possibility of a bomber as far back as 1943 when Specification S.6/43 was issued. This resulted in the production of the Short Sturgeon, a twin-engined naval bomber, which first flew in June 1946. But by that time a change in operational requirements meant that it never saw any operational service, although it was produced in limited numbers as a high-speed target tug. Powered by two Rolls-Royce Merlin engines driving contra-rotating three-bladed propellers, the Sturgeon Mk.1 had a speed of 350 mph and a range of 1,600 nm. With a maximum weight of almost 22,000 lb it was one of the heaviest British naval aircraft of its time. Nevertheless, carrier trials aboard HMS *Illustrious* resulted in very satisfactory reports from the pilots involved who found that carrier operations were much assisted by the contra-rotating propellers, which eliminated torque, and the excellent view from the cockpit located well forward in the nose.

The first Royal Navy aircraft to be capable of delivering nuclear weapons was the Supermarine Scimitar (described in Chapter 6), which became operational in 1958. However, by that time a true nuclear bomber was under development and this was to be one of the most successful British combat aircraft produced since World War II. Following on from the 1951 requirement to deploy tactical nuclear weapons, mainly in the anti-ship role, the Admiralty issued Specification NA39 in the following year for an aircraft capable of delivering such weapons. At the time all the major British aircraft manufacturers presented proposals for what promised to be a very lucrative contract. The winner was Blackburn whose tender was accepted in July 1955 and no fewer than twenty prototypes and pre-production aircraft were ordered by the Ministry of Supply. The Blackburn company's designation was B-103 but the highly secret aircraft quickly became known by its specification number (NA39) and the name Buccaneer had already been adopted when the prototype flew in 30 April 1958. Powered by two de Havilland Gyron Junior turbojets rated at 7,100 lb thrust, the Buccaneer had an extensively area-ruled fuselage and proved to be capable of very high subsonic speeds at very low level, typically 200 feet or below! Although not immediately obvious, its greatest asset was the strength of its airframe, which was tremendously strong and specifically designed to withstand the harsh environment of high-speed, low-level operations. Nuclear or conventional bombs were carried internally in a rotary bomb bay. Although the Buccaneer S.Mk.1 proved to be underpowered, the aircraft was later given a new lease of life when fitted with two more powerful Rolls-Royce Spey engines, although the S.Mk.2 did not fly until 1963. However, this version not only served with the Royal Navy but was also ordered by the RAF in 1968 following the cancellation of the TSR-2 and a subsequent proposal to order American F-111K variable geometry bombers also fell through. The story of Buccaneer development and operational experience falls after the period covered by this book. However, it is relevant to recognise that it came into being as a result of the far-sighted Admiralty specification issued as far back as 1952, and the excellent work of the Blackburn company. (Sadly Blackburn lost its identity in 1963 when it became part of the Hawker Siddeley Group in a wide-scale rationalisation of the British aircraft industry.)

The late 1950s and the 1960s probably represented the golden years for the Royal Navy's Fleet Air Arm in the post-war era. A stable carrier force of four modern carriers (*Eagle, Ark Royal, Victorious* and *Hermes*) was equipped with a range of very capable aircraft such as the Scimitar, Sea Vixen, Buccaneer and Gannet, all of which had their origins in the period covered by this book. Subsequently, rising costs, political decisions and a gradual withdrawal of British forces from the Middle and Far East, resulted in a rundown of the carrier fleet to the extent that conventional fixed-wing carrier flying finally ceased with the de-commissioning of HMS *Ark Royal* in 1978. Since then the Fleet Air Arm has had a reprieve with the operation of Sea Harriers from the three Invincible class carriers and today there is the prospect of two new super carriers in the 60,000-ton class entering service around 2015. However, these are not immune to changes in policy and it remains to be seen if they will ever be completed.

Appendix I

ASW AND AEW

The bulk of this book covers the evolution of frontline naval combat aircraft such as fighters and attack aircraft as they made the transition from piston engine to jet power, and with the changes that were necessary to the ships in order that they could operate the new types coming into service. However, other aircraft were needed to carry out less glamorous but nevertheless essential roles, notably anti-submarine warfare (ASW) and Airborne Early Warning (AEW). In general these were all propeller-driven aircraft and normally did not require the shipboard modifications needed by the jets, although they did benefit from such advances as the angled deck and mirror landing sight. Brief details of these aircraft are given below.

ASW – THE ROYAL NAVY

At the end of World War II, the Royal Navy faced a shortage of modern ASW aircraft. The ancient Swordfish was obsolete and withdrawn from service while the most suitable aircraft, the Grumman Avenger, had to be returned to the US under Lend-Lease agreements. Consequently, a series of stopgap measures were put in hand until a new ASW aircraft could be developed. These included bringing back into service a dozen Fairey Barracuda Mk.IIIs, modified to act as anti-submarine patrol aircraft, in late 1947 to equip 815 Squadron, which retained them until 1953. Their replacement was actually the Grumman Avenger of which 180 were supplied to the Royal Navy from 1953 onwards under the MDAP scheme to boost NATO strength. These were basically similar to the TBM-3E as used by the US Navy (see Chapter 2) but were designated Avenger AS.4 and AS.5 in Royal Navy service.

Up to the introduction of the Avengers, the most numerous carrier-based ASW aircraft was the Fairey Firefly in its AS.5 and AS.6 versions, which are described in Chapter 2. However, in an effort to provide a more mission-orientated aircraft a new version was developed as the Firefly AS.7. The most noticeable change was in the engine installation, which now featured a chin-mounted annular radiator instead of the underwing radiators on the earlier versions. To cope with the increased workload, a third crew member was carried and the two observers/radar operators sat in the rear cockpit, which was covered with a large, bulged clear-view canopy. The addition of this and the chin radiator required a substantial increase in fin area to maintain directional stability. A modified wing plan-form of increased span was incorporated and, although capable of being fitted with underwing hardpoints to carry offensive ordnance, the AS.7 was intended to be used only in the search role with surface ships or other aircraft being directed onto any submarines located. The prototype AS.7 flew on 22 May 1951 and was followed by 150 production aircraft, although only two frontline squadrons were equipped with the type. In service the handling characteristics of the Firefly AS.7 were markedly inferior to the earlier Mks.4/5/6 and consequently most of those built were produced as trainers (Firefly T.Mk.7) and used by second line squadrons for observer training. The poor per-formance of this aircraft was one reason why the Avengers had to be brought in from America, plugging the gap until a completely new ASW aircraft became available.

This was to be another Fairey product, the turboprop-powered Gannet AS.1. Originally known as the Fairey Type Q, it was produced to Specification GR.17/45 issued in 1945, call-ing for a two-seat anti-submarine and strike aircraft. The design was based around an Armstrong Siddeley Double Mamba turbo-prop, which, as the name implied, was actually two 1,000-shp Mamba engines coupled to drive a contra-rotating propeller assembly through a common gearbox. As well as producing the necessary power output, this arrangement had the advantage that one half of the power unit could be shut down in flight to save fuel and extend range and endurance. The first of two prototypes flew in September 1949 but a third prototype flown on 10 May 1951 featured provision for a third crew member in a separate cockpit in the rear half of the fuselage. Although a large aircraft, the Gannet was of a clean design. However, it had a rather portly appearance due to the fact that the weapon load was carried in an internal bomb bay and the crew sat high up above the engine in the nose. The contra-rotating propellers and tricycle undercarriage, coupled with the forward cockpit being set high to give an excellent view, meant that the Gannet was a first-class carrier aircraft and few problems were encountered in the initial deck landing trials by one of the two-seat proto-types aboard HMS *Illustrious* in June 1950 – in fact, this was the first ever carrier landing by a turboprop-powered aircraft. A complex double-jointed wing-folding system ensured that the aircraft would fit in the hangars of current British carriers. An ASV search radar antenna was housed in a retractable dome under the rear fuselage and this meant that the Gannet was able to carry out the complete ASW mission, including both the search and attack roles.

In March 1951 the Gannet was one of the aircraft to be allocated Super Priority status and was ordered in substantial numbers. After some delays while a number of handling prob-lems were sorted out, the first Gannets reached the Royal Navy in April 1954 and three operational squadrons were formed by mid 1955, including 826 Squadron aboard HMS *Eagle* and 824 Squadron aboard HMS *Ark Royal*. Subsequently, the Gannet served with a dozen frontline squadrons and was exported to Australia, West Germany and Indonesia. An improved Gannet AS.4 flew in April 1956 offering a more powerful version of the Double Mamba as well as other detail improvements. The Gannet served as a frontline ASW aircraft until withdrawn between 1958 and 1960 as helicopters gradually took over the ASW role.

Although the Gannet met the requirements of Specification GR.17/45, it had a number of competitors. One of these was the turboprop version of the Short Sturgeon, which used two single Mambas and carried the radar in a prominent fixed chin mounting under the nose. Designated Short SB.3, the prototype flew in August 1950 but whereas the original Sturgeon had been a delightful aircraft to fly, the SB.3 proved something of a handful, particularly in asymmetric flight with one engine failed – a problem that was neatly avoided by the Gannet's Double Mamba installation. Consequently, little development work was undertaken and the two prototypes were scrapped in 1951.

The Gannet's other competitor came from the Blackburn stable in the form of the B-54 (or YA.5 under the then current SBAC designation system). Superficially this resembled the Gannet, being similar in size and adopting the same layout except that the pilot and observer were set further back, seated over rather than ahead off the wing. Like the Gannet, an enclosed weapons bay was set into the lower fuselage and a retractable radar dome was positioned behind it. Power was to be provided by a Napier Double Naiad turboprop, similar in concept and power output to the Double Mamba. However, development of this engine was cancelled leaving the Blackburn team in limbo. It was decided to complete three prototypes powered by Rolls-Royce Griffon piston engines and these were designated YA.7, the first flying in September 1949. When the Admiralty requirement was altered to include a third crew member, the second prototype was modified to incorporate this and also had a revised wing planform and taller fin and rudder assembly. In this form the aircraft became the YA.8 and flew in May 1950, carrying out carrier trials during the following June. Finally, the third prototype was fitted with a Double Mamba turboprop and this represented the definitive version under the designation YB.1 (company designation B-88). First flown in July 1950, it was unsuccessful in comparative trials with the Fairey Gannet and no production orders were forthcoming.

The Gannet might well have had a stablemate in the shape of the Short Seamew. This aircraft was produced in response to a 1951 Naval Staff requirement for a simple and rugged ASW aircraft, in effect a modern version of the old Swordfish, which could be deployed on smaller carriers or from short landing strips ashore. The Seamew first flew on 23 August 1953 and was ordered into production, both for the Royal Navy and RAF Coastal Command. Powered by a single Mamba turboprop, the Seamew was a simple and rugged design with a tailwheel under-carriage necessitated by the decision to place the fixed radome under the fuselage just forward of the wing. A crew of two was carried, with the pilot and observer sitting well forward under a double canopy. Its performance was not startling but the main requirements of a four-hour patrol duration and carriage of depth charges and sonobuoys in an enclosed weapons bay were met. The test programme revealed a number of handling deficiencies that were being rectified when the whole programme was cancelled in 1957 under the Defence White Paper that year. By that time the Royal Navy had already received seven production aircraft and these were subsequently scrapped.

ASW – THE US NAVY

For the US Navy the development of ASW aircraft was not accorded such a high priority as they had numerous Grumman Avengers, which were initially quite suitable for the task and could operate from the smaller Escort Carriers that the US Navy retained for ASW use. However, the increasing amount of electronic equipment that needed to be carried, as well as the increasing use of large weapons such as homing torpedoes meant that a larger aircraft was desirable. In fact, the US Navy avoided an increase in size by splitting up the ASW task into two separate functions, the search or hunting role and the attack or killer role.

The aircraft to carry out this role actually evolved from a Grumman project started in 1944 to design a replacement for the TBF/TBM Avenger. A similar mid wing layout was adopted and the crew of two sat side by side under a single glazed canopy. Power was provided by a single 2,300 hp R-2800-46 Double Wasp radial piston engine but this was supplemented by a single Westinghouse 19XB turbojet in the tail. Prototypes were ordered in February 1945 under the designation XTB3F-1 and the first of these flew on 19 December 1945. However, flight tests revealed that boost provided by the small jet engine was not as much as had been hoped and it was removed, although this did leave additional load-carrying capacity and the design was recast in the ASW role as the Grumman AF-2 Guardian. This had a more powerful 2,400 hp R-2800-48W radial engine and was actually produced in two complementary versions. The AF-2S was a pure weapons carrier, although a short range

APS-30 radar was carried in a pod under the starboard wing, while the AF-2W carried two additional crew members in the rear fuselage and mounted an AN/APS-20 long-range search radar in an under fuselage radome. The Guardian was built in some numbers, with 193 AF-2S and 153 AF-2Ws delivered, along with another forty AF-3S, which was equipped with a Magnetic Anomaly Detector (MAD) in a retractable boom extending aft of the tail assembly. Guardians remained in service until 1957 when the last reserve units relinquished their aircraft.

The Guardian's replacement came from the same Grumman stable and resulted from an attempt to combine the hunter/killer roles into a single airframe. Naturally, this led to a much larger aircraft and the result was the high-winged twin-engined Grumman S2F-1 Tracker. Following a development contract being awarded in 1950, the first Tracker flew on 4 December 1952 and production examples were delivered to a test and evaluation squadron (VX-1) at NAS Key West in July 1953. The first operational squadron (VS-26) was formed in February 1954 and the Tracker was subsequently deployed aboard all types of operational carriers, including some escort carriers (CVEs) utilised in the ASW role. Although a relatively large aircraft, the Tracker could be accommodated aboard the smaller carriers due to its short overall length and compact dimensions with wings folded. A total of 1,120 Trackers were built by Grumman and a further 100 were built by de Havilland Canada for the Canadian Navy as the CS2F-1. The Tracker was also operated by several other navies, including those of Australia, Brazil, Argentina and the Netherlands, which all deployed them aboard carriers, while Italy, Japan, Korea and Thailand were among those who operated Trackers from shore bases. The Tracker remained in frontline US Navy service until the mid 1970s, by which time it was being replaced by the jet-powered Lockheed S-3 Viking.

AEW – ROYAL AND US NAVIES
It was the US Navy that instigated the concept of a naval airborne early warning aircraft as a result of the damage caused

by Japanese kamikaze attacks in the closing stages of World War II. It was vital to detect these low-flying raids as far out as possible and the destroyer picket ships used for this purpose became prime targets themselves. The development of a radar system that could be carried aboard an aircraft for this purpose was code-named Project Cadillac and began in February 1944. The resulting AN/APS-20 radar was fitted to a modified TBM Avenger in August 1944 and delivery of radar-equipped Avenger TBM-3Ws began in May 1945, although operational aircraft were just too late to see service in World War II. Nevertheless a large number of TBM-3s were modified and by the height of the Korean War in early 1953 over 150 were in service. Apart from the radome, the other changes were the removal of all armament including the dorsal turret, a modified rear cockpit to house the radar operator, and two large finlets on the tailplane to counteract the adverse effect of the radome. The accommodation was extremely cramped and the single radar operator could easily be overloaded in a busy action scenario.

In the early 1950s the TBM-3Ws were gradually replaced by AEW versions of the AD Skyraider (described in Chapter 2). Forty-five AD-4Ws were also supplied to the Royal Navy from 1951 onwards, remaining in service until 1960. The size of the Skyraider allowed two radar operators to be accommodated in the rear fuselage, which much improved the service these aircraft were able to offer. Eventually, the Skyraider AEW version was replaced in US Navy service by the Grumman WF Tracer, which was a development of the S2F Tracker in which an enormous radome was fitted atop the fuselage, necessitating the adoption of a wide-span tailplane with twin fins. However, this did not fly until March 1958 and only became operational in 1960, although it subsequently formed the basis for the turboprop Hawkeye, which is still in production today.

The Royal Navy's replacement for the Skyraider AEW.1 was the homegrown Fairey Gannet AEW.3. This was a development of the basic Gannet design in which a large radome was carried under the fuselage centre section, replacing the weapons bay that was no longer required. The pilot's cockpit remained in the original position but the rear cockpits were deleted and two radar operators were accommodated in a cabin within the fuselage. However, the Gannet AEW.3 did not fly until August 1958. It then became operational in July 1960 when aircraft of C flight, 849 Squadron, embarked on the recently commissioned carrier HMS *Hermes*.

Appendix II

FRENCH NAVAL AVIATION

Inevitably this book has concentrated on the experience of the US and Royal Navies as they were the front runners in the development of carrier-based aviation during the period in question. A number of other navies operated aircraft carriers, noticeably those of Australia, Canada and the Netherlands, but in each case these used ex-British or American carriers and obtained their aircraft from the same sources (although the Netherlands did produce the Sea Fury under licence). However, the French Navy was in a slightly different category. Although prior to World War II they had made little progress and had only commissioned one aircraft carrier (*Bearn*), plans existed to build two new 18,000-ton carriers. The *Bearn* proved unsuitable for operations during the War and was eventually used as an aircraft transport, although it did serve briefly in Indo-China in 1945/6 equipped with a motley collection of aircraft, which included captured Japanese seaplanes and Piper L-4 Cub observation aircraft. She was paid off in July 1946 at Toulon.

Towards the end of the War the ex-Royal Navy escort carrier HMS *Biter* was transferred to the French Navy as the *Dixmude*. A more substantial contribution was the transfer of the modern light fleet carrier HMS *Colossus* in March 1947 when she was renamed *Arromanches*. Finally, two ex-US Navy light fleet carriers (*Langley* and *Belleau Wood*) were transferred in 1951 and 1953 respectively, becoming the *La Fayette* and *Bois Belleau*. To equip these carriers Britain supplied forty-eight Seafire Mk.IIIs and later fifteen Griffon-engined Seafire Mk.XVs. In addition, significant numbers of ex-American aircraft were supplied, including Dauntless, Helldivers, Avengers, Hellcats and Corsairs – the latter including ninety-four new-build

F4U-7s, which were the last to be built by Chance Vought. All of these aircraft saw substantial action in the post-war decade, flying support missions in Indo-China before France was forced to withdraw in 1954, and also later in support of French troops in the Algerian War. France also acquired its first jet aircraft in the form of licence-built versions of the British de Havilland Sea Venom. This was built in both single and two-seat versions and given the name Aquilon.

However, France is particularly interesting in the context of this book in that it was the only country apart from Britain and America to produce naval aircraft of indigenous design, although these only flew in prototype form and none entered service. A number of projects were initiated in the immediate post-war era and in many cases French engineers were able to draw on experience gained when they had been forced to work on German projects. In particular, some had experience of the Jumo 004 jet engine that had powered the Me.262 jet fighter. One of the first to fly was the Arsenal VG70, which was a single-seat research aircraft powered by a Jumo 004 turbojet. It also incorporated advanced features such as a swept wing and a tricycle undercarriage (at a time when Britain was producing the straight-winged Attacker with a tailwheel undercarriage). An unusual feature of the design was the ventral under fuselage air intake (rather like the modern F-16). The first flight was delayed until June 1948 and a maximum speed of 559 mph was subsequently attained. However, the programme was plagued by the unreliability of the Jumo turbojet but it did lead to the development of the VG90 intended to meet a French Navy requirement for a naval fighter. The general configuration was retained but the powerplant was a Rolls-Royce Nene with twin lateral intakes beneath the shoulder-mounted swept wing. Two prototypes were built and the first flight was on 27 September 1949, although this aircraft was destroyed in an accident in May 1950. Tests revealed a maximum speed of 596 mph and an initial rate of climb of 4,530 feet/min.

A second jet fighter designed to the same *Aeronavale* speci-fication was the Nord 2200, which first flew on 19 December 1949. This was an attractive swept wing aircraft vaguely reminiscent of the American F-86 Sabre and was also powered by a Rolls-Royce Nene turbojet. The maximum speed was 581 mph, the rate of climb around 4,500 feet/min and the Nord 2200 had a very respectable endurance of one hour thirty minutes on internal fuel. The armament was either three 20 mm or 30 mm cannon and two 1,100 lb (500 kg) bombs could be carried underwing. No production order was forthcoming but the aircraft was retained for extensive testing and evaluation, and valuable experience relating to the operation of swept wing aircraft was obtained.

A third fighter prototype was the Aerocentre NC.1080, which flew in July 1949 and bore a passing resemblance to the Supermarine 525. Again powered by the ubiquitous 5,000 lb thrust Rolls-Royce Nene turbojet, the NC.1080 was a graceful swept wing aircraft with a tricycle undercarriage and a straight tailplane set midway up the swept tail fin. The top speed was 584 mph but the rate of climb was just under 4,000 feet/min, although the service ceiling was in excess of 43,500 feet. This promising aircraft was destroyed in an accident that brought a halt to further development and the decision to build the Sea Venom under licence effectively ended the other programmes at the time.

However, one aircraft produced in this period did eventually see operational service, but not in its original form. This was the Breguet 960 Vultur, which was designed as an attack aircraft and was powered by a combination of a 1,000 shp Armstrong Siddeley Mamba turboprop in the nose and a licence-built Hispano-Suiza/Rolls-Royce Nene in the rear fuselage. Development began in late 1947 and the prototype Vultur flew on 3 August 1951. Carrying a crew of two, it featured a straight wing but with a sharply tapered leading edge and small tip tanks. In this form it could reach speeds of almost 550 mph on both engines but could cruise at 230 mph on the Mamba turboprop, giving it an endurance of over four hours. The offensive load consisted of a single 2,200 lb (1,000 kg) bomb on the under fuselage centreline, and eight rockets on underwing rails. Three prototypes were flown but already the *Aeronavale* requirement was changing and the strike role was eventually to be carried out by the all-jet Dassault Etendard IVM, which first flew in 1958 and entered service in 1961. However, it was decided to use the Vultur as a basis for a new anti-submarine aircraft and the third prototype Vultur was modified to test some of the aerodynamic features of the new project, which became the Breguet Type 965 Alize. Although this retained the basic layout of the Vultur, the mixed power concept was abandoned in favour of a single Rolls-Royce Dart turboprop and the rest of the airframe underwent a considerable redesign so that the prototype did not fly until March 1956. Subsequently, a total of ninety-two Alizes were built, of which twelve were ordered by India who later received a further twelve ex-*Aeronavale* aircraft. The Alize entered operational service in 1959 and was in many ways very similar to the British Gannet, although its career was much longer, the last examples only being retired in September 2000. When the Alize entered service the French Navy was building two 30,000-ton carriers (*Clemenceau* and *Foch*), which commissioned in the early 1960s. By the time it retired these two ships were in the process of being replaced by the nuclear-powered *Charles de Gaulle* equipped with supersonic Rafale jet fighters and Super Etendard strike aircraft – putting France into the front rank of today's naval aviation.

Appendix III

AIRCRAFT SPECIFICATIONS – 1945 TO 1955

This appendix gives details of most of the British and American naval aircraft developed in the period 1945 to 1955. Within each section, the aircraft are listed in order of their first flight to give a sense of the rate of progress made at the time. In most cases, the performance figures given are intended only as a guide as parameters such as maximum speed and range may well have been reduced in operational configurations. Dimensions are given to the nearest inch.

ROYAL NAVY PROPELLOR-DRIVEN AIRCRAFT

Fighters/Fighter-bombers

Fairey Firefly FR.Mk.4
Two-seat piston engine fighter reconnaissance

Power:	One Rolls-Royce Griffon 74, 2,245 hp
Armament:	Four 20 mm Hispano cannon; provision for two 1,000 lb bombs or sixteen 60 lb warhead rocket projectiles
Max speed:	367 mph at 14,000 ft
Range:	760 miles (1,335 miles with two 90-gal auxiliary tanks)
Climb:	7 min 9 sec to 10,000 ft
Service ceiling:	31,900 ft
Span:	41 ft 2 in (13 ft 6 in folded)
Length:	38 ft
Height:	12 ft 4 in

Empty weight: 9,674 lb
Loaded weight: 13,479 lb (15,615 lb max overload)

First flew 1944. Total of 591 Mks 4/5/6 delivered. In service
1947–58.

Supermarine Seafire F.XVII
Single-seat piston engine fighter

Power: One Rolls-Royce Griffon VI, 1,850 hp
Armament: Two 20 mm Hispano cannon; four
                           0.303-in machine-guns; one 500 lb or two
                           250 lb bombs; provision for eight 60 lb
                           rocket projectiles.
Max speed: 387 mph at 13,500 ft
Range: 435 miles
Max rate of climb: 4,600 ft/min at sea level
Service ceiling: 35,200 ft
Span: 36 ft 10 in
Length: 32 ft 3 in
Height: 10 ft 8 in
Empty weight: 6,200 lb
Loaded weight: 8,000 lb

232 built under sub contract by Westland and Cunliffe-Owen.
In service 1945–54.

Hawker Sea Fury FB.Mk.11
Single-seat piston engine fighter-bomber

Power: One Bristol Centaurus 18 air-cooled
                           radial engine, 2,480 hp
Armament: Four 20 mm Hispano cannon; provision
                           for two 1,000 lb bombs or twelve 3-in
                           rocket projectiles
Max speed: 460 mph at 18,000 ft
Range: 700 miles (1,040 miles with two 90-gal
                           drop tanks)
Climb: 10.8 min to 30,000 ft
Service ceiling: 35,800 ft
Span: 38 ft 5 in

Length:	34 ft 8 in
Height:	15 ft 10 in
Empty weight:	9,240 lb
Loaded weight:	12,500 lb

First flew February 1945. 615 delivered to the Royal Navy including 50 F.Mk.10s and 60 T.Mk.20s. Also served with Australian and Canadian Navies.

De Havilland DH.103 Sea Hornet F.Mk.20
Single-seat twin piston engine fighter-bomber

Power:	Two 2,030 hp Rolls-Royce Merlin 130/131 or 134/135
Armament:	Four 20 mm Hispano cannon; provision for two 1,000 lb bombs
Max speed:	467 mph at 22,000 ft
Range:	1,500 miles
Max rate of climb:	4,000 ft/min
Service ceiling:	35,000 ft
Span:	45 ft 0 in
Length:	36 ft 9 in
Height:	13 ft 0 in
Empty weight:	13,300 lb
Loaded weight:	18,530 lb

First flew 19 April 1945. 198 of all versions (F.Mk.20, NF.Mk.21, PR.Mk.22) delivered to the Royal Navy. In service 1946–55.

Supermarine Seafang 32
Single-seat piston engine fighter

Power:	One Rolls-Royce Griffon 89, 2,375 hp
Armament:	Four 20 mm Hispano cannon; provision for two 1,000 lb bombs or four 300 lb rocket projectiles
Max speed:	475 mph at 21,000 ft
Range:	393 miles
Max rate of climb:	4,630 ft/min
Service ceiling:	41,000 ft
Span:	35 ft

Length:	34 ft 1 in
Height:	12 ft 6 in
Empty weight:	8,000 lb
Loaded weight:	10,450 lb

First flew January 1946. Ten Seafang 31s and eight Seafang 32s completed. Not used operationally.

Supermarine Seafire FR.47
Single-seat piston engine fighter

Power:	One Rolls-Royce Griffon 87 or 88, 2,350 hp
Armament:	Four 20 mm Hispano cannon; three 500 lb or 250 lb bombs; provision for eight 60 lb rocket projectiles or two Mk.IX depth charges
Max speed:	451 mph at 20,000 ft
Range:	405 miles (1,000 miles with additional tanks)
Max rate of climb:	4,800 ft/min at sea level
Service ceiling:	43,100 ft
Span:	36 ft 11 in
Length:	34 ft 4 in
Height:	12 ft 9 in
Empty weight:	7,625 lb
Loaded weight:	10,200 lb (12,750 lb max overload)

90 built by Vickers at South Marston. In service 1947–52.

Attack and Strike Aircraft

Blackburn Firebrand TF.Mk.5
Single-seat piston engine torpedo bomber

Power:	One Bristol Centaurus IX air-cooled radial engine, 2,520 hp
Armament:	Four 20 mm Hispano cannon; provision for one 1,850 lb 18-in torpedo, or two 1,000 lb bombs or sixteen 3-in rocket projectiles

Max speed:	340 mph at 13,000 ft (320 mph with torpedo)
Range:	740 miles
Initial rate of climb:	2,500 ft/min
Service ceiling:	31,000 ft
Span:	51 ft 3 in (folded 16 ft 10 in)
Length:	38 ft 9 in
Height:	13 ft 3 in
Empty weight:	11,835 lb
Loaded weight:	17,500 lb

First flew May 1945 (TF.Mk.4). 220 Firebrands of all variants built. In service 1945–53

Fairey Spearfish
Two-seat piston engine naval strike aircraft

Power:	One Bristol Centaurus air-cooled radial engine, 2,600 hp
Armament:	Two fixed wing-mounted Browning 0.5-in machine-guns; two 0.5-in machine-guns in a remotely controlled Frazer-Nash 95 barbette; internal bomb bay capable of accommodating one torpedo, or one 2,000 lb or four 500 lb bombs, or four depth charges; sixteen 60 lb rockets on underwing zero-length launchers
Max speed:	292 mph at 14,000 ft
Range:	1,036 miles
Climb:	7 min 45 sec to 10,000 ft
Service ceiling:	25,000 ft
Span:	60 ft 3 in (20 ft folded)
Length:	44 ft 7 in
Height:	13 ft 6 in
Empty weights:	15,200 lb
Loaded weight:	21,642 lb (22,083 lb max overload)

First flew 5 July 1945. Two prototypes tested. Third aircraft completed but not flown.

Westland Wyvern S.4
Single-seat turboprop torpedo strike aircraft

Power:	One Armstrong Siddeley Python A.S.P.3., 4,110 ehp
Armament:	Four 20 mm Hispano cannon; provision for one 18-in torpedo, or three 1,000 lb bombs or sixteen 60 lb warhead rocket projectiles
Max speed:	383 mph at sea level
Range:	704 miles
Initial rate of climb:	2,350 ft/min
Service ceiling:	28,000 ft
Span:	44 ft (20 ft folded)
Length:	42 ft 3 in
Height:	15 ft 8 in
Empty weight:	15,600 lb
Loaded weight:	21,200 lb (24,500 lb max overload)

Prototype TF.1 flew 12 December 1946. 97 production S.4s delivered. In service 1953–7.

Blackburn B-48 Firecrest
Single-seat piston engine torpedo bomber

Power:	One Bristol Centaurus 59 air-cooled radial engine, 2,475 hp
Armament:	Provision for one 1,850 lb 18-in torpedo, or two 500 lb bombs or twelve 3-in rocket projectiles or two 0.5-in machine-guns in underwing pods
Max speed:	380 mph at 19,000 ft
Range:	900 miles
Initial rate of climb:	2,500 ft/min
Service ceiling:	31,600 ft
Span:	44 ft 1 in (folded 18 ft 0 in)
Length:	39 ft 3 in
Height:	14 ft 6 in
Empty weight:	10,513 lb
Loaded weight:	15,280 lb

First flew April 1947. Three prototypes, of which only two were flown.

Anti-submarine

Fairey Gannet AS.1
Three-seat turboprop torpedo anti-submarine patrol aircraft

Power:	One Armstrong Siddeley Double Mamba 100 turboprop, 2,950 ehp
Armament:	Two homing torpedoes and three depth charges, or 2,000 lb of bombs in internal bomb bay; sixteen 60 lb rocket projectiles on underwing racks
Max speed:	310 mph at sea level
Range:	662 miles
Initial rate of climb:	1,900 ft/min
Service ceiling:	25,000 ft
Span:	54 ft 4 in
Length:	43 ft
Height:	13 ft 8 in
Empty weight:	15,069 lb
Loaded weight:	19,600 lb

Prototype flew 10 September 1949. 348 Gannets of all versions delivered, including some for export and 44 AEW.3s. In RN service 1954–60 (ASW variants).

Fairey Firefly AS.Mk.7
Three-seat piston engine anti-submarine patrol

Power:	One Rolls-Royce Griffon 59, 1,965 hp
Armament:	No offensive armament carried
Max speed:	300 mph at 10,750 ft
Range:	860 miles
Climb:	1,550 ft/min
Service ceiling:	25,500 ft
Span:	44 ft 6in
Length:	38 ft 3 in
Height:	13 ft 3 in

Empty weight: 11,016 lb
Loaded weight: 13,970 lb

First flew 22 May 1951. Total of 151 delivered. In service 1952–7.

UNITED STATES PROPELLER DRIVEN AIRCRAFT

Fighters/Fighter-bombers

Grumman F7F-4N
Two-seat twin-engined piston engine fighter

Power:	Two Pratt & Whitney R-2800-34W air-cooled radial engines, 2,100 hp each
Armament:	Four 20 mm cannon; provision for one 2000 lb bomb or one Mk.13 torpedo on a centreline rack, or two 1,000 lb bombs on inboard underwing racks, or eight 250 lb bombs or 5-in HVAR rockets on outboard underwing racks
Max speed:	436 mph at 24,500 ft
Range:	1,360 miles
Initial rate of climb:	4,100 ft/min
Service ceiling:	37,600 ft
Span:	51 ft 6 in (folded 32 ft 2 in)
Length:	46 ft 9 in
Height:	16 ft 4 in
Empty weight:	16,954 lb
Loaded weight:	21,960 lb

Prototype XF7F-1 first flew 3 November 1943. A total of 364 Tigercats were delivered, including 13 F7F-4Ns, the final production version. In service (all combat variants) 1944–53.

Grumman F6F-5 Hellcat

Single-seat piston engine fighter/fighter-bomber

Power:	Pratt & Whitney R-2800-10W air-cooled radial engine, 2,000 hp
Armament:	Six 0.5-in machine-guns; provision for two 1,000 lb bombs or six 5-in rockets
Max speed:	386 mph at 17,300 ft

Range:	1,040 miles
Initial rate of climb:	3,410 ft/min
Service ceiling:	37,300 ft
Span:	42 ft 10 in
Length:	33 ft 7 in
Height:	13 ft 1 in
Empty weight:	9,042 lb
Loaded weight:	11,381 lb

First flew 4 April 1944. 6,436 F6F-5s built and total of 12,272 Hellcats of all versions. In service (all versions) 1943–54.

Grumman F8F-1 Bearcat
Single-seat piston engine interceptor fighter

Power:	Pratt & Whitney R-2800-34W air-cooled radial engine, 2,100 hp
Armament:	Four 0.5-in machine-guns (four 20 mm cannon in F8F-1B); provision for two 1,000 lb bombs
Max speed:	434 mph at 20,000 ft
Range:	1,105 miles
Initial rate of climb:	4,570 ft/min
Service ceiling:	38,700 ft
Span:	35 ft 10 in
Length:	28 ft 3 in
Height:	13 ft 10 in
Empty weights:	7,070 lb
Loaded weight:	12,950 lb

First flew 21 August 1944. 1,263 of all versions produced. In service (all versions) 1945–53.

Boeing XF8B-1
Single-seat piston engine long-range fighter and attack aircraft

Power:	Pratt & Whitney R-4360-10 air-cooled radial engine, 3,000 hp
Armament:	Six 20 mm cannon or 0.5-in machine-guns; four 500 lb or two 1,600 lb bombs or two torpedoes in internal bomb bay;

	two 1,600/1,000/500 lb bombs on under-wing racks
Max speed:	432 mph at 26,900 ft
Range:	1,305 miles (maximum over 3,000 miles with auxiliary tanks)
Initial rate of climb:	2,880 ft/min
Service ceiling:	37,500 ft
Span:	54 ft 0 in
Length:	43 ft 3 in
Height:	16 ft 3 in
Empty weight:	13,519 lb
Loaded weight:	21,691 lb

First flew 27 November 1944. Three prototypes only.

Goodyear F2G-2
Single-seat piston engine low-altitude fighter-bomber

Power:	Pratt & Whitney R-4360-4 air-cooled radial engine, 3,000 hp
Armament:	Four 0.5-in machine-guns; provision for two 1,600 or 1,000 lb bombs, or eight 5-in rockets
Max speed:	431 mph at 16,400 ft
Range:	1,190 miles
Initial rate of climb:	4,400 ft/min
Service ceiling:	38,800 ft
Span:	41 ft 0 in
Length:	33 ft 9 in
Height:	16 ft 1 in
Empty weight:	10,249 lb
Loaded weight:	15,442 lb (max)

First flight early 1945. Production totalled five F2G-1s and five F2G-2s.

Vought F4U-5 Corsair
Single-seat piston engine fighter-bomber

| | |
| Power: | Pratt & Whitney R-2800-32W air-cooled radial engine, 2,300 hp |

Armament:	Four 20 mm cannon; provision for two 1,000 lb bombs
Max speed:	470 mph at 26,800 ft
Range:	1,120 miles
Initial rate of climb:	3,780 ft/min
Service ceiling:	41,400 ft
Span:	41 ft 0 in
Length:	33 ft 6 in
Height:	14 ft 9 in
Empty weight:	9,683 lb
Loaded weight:	14,106 lb

First F4U-5 flew 1946. Production consisted of 223 F4U-5s, 315 F-4U-5N night fighters and 30 F4U-5Ps. In service (all Corsair variants) 1942–57.

Attack and Strike Aircraft

Curtiss SB2C-4 Helldiver
Two-seat single piston engine dive bomber

Power:	One Wright R-3350-14 air-cooled radial engine, 2,300 hp
Armament:	Two wing-mounted 20 mm cannon; two 1,600 lb bombs in internal bomb bay, plus two 500 lb bombs on underwing racks or two 2,100 lb torpedoes
Max speed:	344 mph at 16,100 ft
Range:	1,480 miles (2,140 miles max)
Initial rate of climb:	1,650 ft/min
Service ceiling:	23,600 ft
Span:	45 ft
Length:	38 ft 7 in
Height:	16 ft 7 in
Empty weight:	12,900 lb
Loaded weight:	18,140 lb (19,000 lb maximum)

Prototype XSB2C-1 first flew on 18 December 1940. A total of 5,399 Helldivers built, including 1,294 produced in Canada. In service (all combat variants – US Navy) 1943–9.

Grumman (Eastern Aircraft) TBM-3E Avenger
Three-seat single-engine torpedo bomber

Power:	One Wright R-2600-20 air-cooled radial engine, 1,900 hp
Armament:	Two wing-mounted 0.5-in machine-guns; one 0.5-in machine-gun in a dorsal turret; one 0.3-in machine-gun in a ventral mounting; one torpedo or 2,000 lb of bombs in internal bomb bay; provision for eight 5-in rockets carried underwing
Max speed:	276 mph at 16,500 ft
Range:	1,010 miles
Initial rate of climb:	2,060 ft/min
Service ceiling:	30,000 ft
Span:	54 ft 2 in (folded 19 ft 0 in)
Length:	41 ft 0 in
Height:	16 ft 5 in
Empty weight:	10,545 lb
Loaded weight:	17,895 lb (max)

Prototype XTBF-1 first flew on 1 August 1941. A total of 9,839 Avengers built by Grumman and Eastern. In service (all combat variants – US Navy) 1942–54.

Douglas BTD-1 Destroyer
Single-seat single piston engine attack bomber

Power:	One Wright R-2600-20 air-cooled radial engine, 1,900 hp
Armament:	Two wing-mounted 20 mm cannon; two 0.3-in machine-guns in rear cockpit; 1,000 lb of bombs in internal bomb bay; two 500 lb bombs or eight 5-in rockets on underwing racks
Max speed:	295 mph at 16,700 ft
Range:	1,165 miles
Initial rate of climb:	1,800 ft/min
Service ceiling:	29,100 ft
Span:	49 ft 9 in
Length:	36 ft 8 in

Height:	13 ft 2 in
Empty weight:	10,545 lb
Loaded weight:	16,616 lb

Developed from two-seat XSB2D-1, which flew on 8 April 1943. First single-seat BTD-1 flew on 5 March 1944. 27 built, not used operationally.

Martin AM-1 Mauler
Single-seat single piston engine attack bomber

Power:	One Pratt & Whitney R-3350-4 radial engine, 2,975 hp
Armament:	Four wing-mounted 20 mm cannon; normally up to 5,000 lb of bombs, torpedoes or rockets on underwing hardpoints (maximum ordnance load of 10,689 lb demonstrated)
Max speed:	367 mph at 11,600 ft
Range:	1,800 miles
Initial rate of climb:	2,780 ft/min
Service ceiling:	30,500 ft
Span:	50 ft
Length:	41 ft 2 in
Height:	16 ft 10 in
Empty weight:	14,500 lb
Loaded weight:	23,386 lb

First flight on 26 August 1944. 151 built. In service 1948–52.

Douglas XTB2D-1 Skypirate
Three/four-seat single piston engine attack bomber

Power:	One Pratt & Whitney XR-4360-8 air-cooled radial engine, 3,000 hp
Armament:	Four wing-mounted 0.5-in machine-guns; two 0.5-in machine-guns in dorsal turret; one 0.5-in machine-gun in ventral mounting; up to 8,400 lb of bombs, torpedoes or depth charges on underwing racks
Max speed:	340 mph at 15,600 ft

Range: 1,250 miles (2,880 miles max)
Initial rate of climb: 1,390 ft/min
Service ceiling: 24,500 ft
Span: 70 ft
Length: 46 ft
Height: 22 ft 7 in
Empty weight: 18,405 lb
Loaded weight: 28,545 lb (34,750 lb max)

First flight on 13 March 1945. Single prototype only, production orders cancelled late 1945.

Douglas AD-4 Skyraider
Single-seat single piston engine attack bomber

Power: One Pratt & Whitney R-3350-26WA radial engine, 2,700 hp
Armament: Two or four wing-mounted 20 mm cannon; normally up to 8,000 lb of bombs, torpedoes or rockets on underwing hardpoints (max ordnance load of 10,500 lb demonstrated)
Max speed: 320 mph at 15,000 ft
Range: 900 miles
Initial rate of climb: 2,980 ft/min
Service ceiling: 23,800 ft
Span: 50 ft
Length: 39 ft 3 in
Height: 15ft 8 in
Empty weight: 11,783 lb
Loaded weight: 24,221 lb (max)

Prototype flew as XBT2D-1 on 18 March 1945. Subsequently redesignated AD. 3,180 delivered bewteen 1945 and 1957. In US Navy service 1946–71.

Kaiser Fleetwings XBTK-1
Single-seat single piston engine attack bomber

Power: One Pratt & Whitney R-2800-34W radial engine, 2,100 hp

Armament:	Two wing-mounted 20 mm cannon; up to 5,000 lb of bombs, or one torpedo and eight 5-in rockets
Max speed:	342 mph at sea level
Range:	1,250 miles
Service ceiling:	33,400 ft
Span:	48 ft 8 in
Length:	38 ft 11 in
Empty weight:	9,959 lb
Loaded weight:	12,728 lb (normal)

First flight in April 1945. Two prototypes flown.

Curtiss XBTC-2
Single-seat single piston engine attack bomber

Power:	One Pratt & Whitney XR-4360-8 air-cooled radial engine, 3,000 hp
Armament:	Four wing-mounted 20 mm cannon; two 1,000 lb bombs or one torpedo
Max speed:	374 mph at 16,000 ft
Range:	1,835 miles
Initial rate of climb:	2,250 ft/min
Service ceiling:	26,200 ft
Span:	50 ft
Length:	39 ft
Height:	12 ft 11 in
Empty weight:	13,410 lb
Loaded weight:	21,660 lb (max)

First flight in July 1946. Two prototypes only.

Lockheed P2V-3 Neptune
Multi-crew twin-engined strategic bomber

Power:	Two Wright R-3350-26W air-cooled radial engines, each 3,200 hp
Armament:	Up to 10,000 lb of bombs including Mk.1 14-kiloton atomic bomb
Max speed:	337 mph at 13,000 ft
Range:	3,930 miles

Initial rate of climb: 1,060 ft/min
Service ceiling: 28,000 ft
Span: 100 ft
Length: 77 ft 10 in
Height: 28 ft 1 in
Empty weight: 34,875 lb
Loaded weight: 64,100 lb (max)

First P2V-3 flew on 6 August 1948. Twelve aircraft converted as P2V-3Cs for operation from carriers used rocket-assisted take-off.

Douglas XA2D-1 Skyshark
Single-seat turboprop engine attack bomber

Power: One Allison XT40-A-2 twin turboprop, 5,100 shp
Armament: Four wing-mounted 20 mm cannon; normally up to 5,500 lb of bombs, torpedoes or rockets on underwing hardpoints
Max speed: 501 mph at 25,000 ft
Range: 2,200 miles (max)
Initial rate of climb: 7,290 ft/min
Service ceiling: 48,100 ft
Span: 50 ft
Length: 41 ft 2 in
Height: 17 ft
Empty weight: 12,944 lb
Loaded weight: 22,966 lb (max)

First flight on 26 May 1950. Two prototypes and ten production aircraft (four not flown). No operational service.

Anti-submarine

Grumman AF-2S Guardian
Three-seat single piston engine anti-submarine patrol aircraft

Power: One Pratt & Whitney R-2800-46W air-cooled radial engine, 2,300 hp

Armament:	One homing torpedo in bomb bay and/or four 500 lb bombs, or four Mk.54 depth charges, or six 5-in HVAR rockets on underwing racks
Max speed:	275 mph at 4,000 ft
Range:	1,140 miles (max)
Initial rate of climb:	2,300 ft/min
Service ceiling:	22,900 ft
Span:	60 ft (24 ft folded)
Length:	42 ft 1 in
Height:	13 ft 2 in
Empty weight:	14,658 lb
Loaded weight:	22,565 lb (max)

First flew in November 1948 as XTB3F-1. AF-2W similar but carried search radar and no armament. 389 Guardians built, including three XTB3F-1 protypes. In service 1950–57.

Grumman S2F-1 Tracker
Four-seat twin piston engine anti-submarine patrol aircraft

Power:	Two Wright R-1820-82 air-cooled radial engines, each 1,525 hp
Armament:	Two homing torpedoes or one Mk.24 mine in ventral bomb bay; either four Mk.19 mines, four Mk.43 torpedoes, four Mk.54 depth charges, or six HVAR 5-in rockets on underwing racks
Max speed:	272 mph at 3,000 ft
Range:	968 miles
Initial rate of climb:	2,330 ft/min
Service ceiling:	22,800 ft
Span:	69 ft 8 in (27 ft 4 in folded)
Length:	42 ft
Height:	16 ft 3 in
Empty weight:	17,355 lb
Loaded weight:	24,408 lb (max)

First flight on 4 December 1952. 1,269 S2Fs produced, including 100 built in Canada. In US Navy service 1954–76 (ASW variants).

VTOL Prototypes

Lockheed XFV-1
Experimental turboprop vertical take-off fighter

Power:	One Allison XT40-A-6 double turboprop, 5,850 ehp
Armament:	Proposals for either four 20 mm cannon or 48 2.75-in FFAR rockets
Max speed:	610 mph at 15,000 ft
Climb:	Up to 30,000 ft in 4.6 min
Service ceiling:	43,700 ft
Span:	25 ft 8 in
Length:	30 ft 9 in
Empty weight:	11,600 lb
Loaded weight:	16,220 lb

Single prototype – first flight in conventional mode on 16 June 1954. No vertical take-offs or landings completed, although inflight hovers achieved. Programme cancelled in June 1955.

Convair XFY-1
Experimental turboprop vertical take-off fighter

Power:	One Allison XT40-A-6 double turboprop, 5,850 ehp
Armament:	Proposals for either four 20 mm cannon or 48 2.75-in FFAR rockets
Estimated Performance with more powerful YT40 engine:	
Max speed:	580 mph at 15,000 ft
Endurance:	70 min
Initial rate of climb:	10,280 ft/min
Service ceiling:	43,300 ft
Span:	30 ft 10 in
Length:	36 ft 10 in
Loaded weight:	c.11,000 lb

Single prototype. First free flight involving vertical take-off and landing, and transition to horizontal flight, made in November 1954. Programme cancelled in 1956.

ROYAL NAVY JET AIRCRAFT

Fighters/Fighter-bombers

De Havilland Sea Vampire F.Mk.20
Single-seat jet fighter

Power:	One de Havilland Goblin 2 turbojet, 3,000 lb st
Armament:	Four 20 mm cannon
Max speed:	526 mph
Range:	1,145 miles (max)
Initial rate of climb:	4,300 ft/min
Service ceiling:	43,000 ft
Span:	38 ft (non-folding wings)
Length:	30 ft 9 in
Height:	8 ft 10 in
Empty weight:	7,623 lb
Loaded weight:	11,970 lb

Prototype Vampire flew on 29 September 1943. 18 production Sea Vampire F.Mk.20s and three F.Mk.21s. In service 1948–53.

Supermarine Attacker F.Mk.1
Single-seat jet fighter

Power:	One Rolls-Royce Nene 3 turbojet, 5,100 lb st
Armament:	Four wing-mounted 20 mm cannon
Max speed:	590 mph at sea level
Range:	1,190 miles (with ventral tank)
Initial rate of climb:	6,350 ft/min
Service ceiling:	45,000 ft
Span:	36 ft 11 in
Length:	37 ft 6 in
Height:	9 ft 11 in
Empty weight:	8,434 lb
Loaded weight:	11,500 lb

Prototype flew on 27 July 1946. 3 prototypes and 145 production aircaft (F.1, FB.1 and FB.2) In service 1951–7.

Hawker Sea Hawk F.Mk.1
Single-seat jet fighter

Power:	One Rolls-Royce Nene 4 turbojet, 5,000 lb st
Armament:	Four 20 mm cannon
Max speed:	560 mph
Range:	1,000 miles
Initial rate of climb:	5,000 ft/min
Service ceiling:	44,000 ft
Span:	39 ft (13 ft 4 in folded)
Length:	39 ft 10 in
Height:	8 ft 9 in
Empty weight:	9,190 lb
Loaded weight:	13,220 lb

Prototype Hawker P.1040 first flew on 2 September 1947. 525 Seahawks of all variants produced. In Royal Navy service 1953–60.

Hawker P.1052
Single-seat swept wing research aircraft

Power:	One Rolls-Royce Nene 2 turbojet, 5,000 lb st
Armament:	None fitted
Max speed:	680 mph at sea level
Climb:	9 min 30 sec to 35,000 ft
Service ceiling:	45,500 ft
Span:	31 ft 6 in
Length:	39 ft 7 in
Height:	10 ft 6 in
Empty weight:	9,450 lb
Loaded weight:	13,448 lb

Swept wing development of P.1040 Seahawk. First of two prototypes flew on 19 November 1948. Carrier trials carried out but not developed for service use.

Supermarine Type 510
Single-seat swept wing jet fighter

Power:	One Rolls-Royce Nene 2 turbojet, 5,000 lb st
Armament:	None fitted

Max speed:	635 mph at 15,000 ft
Service ceiling:	40,000 ft
Span:	31 ft 8 in
Length:	38 ft 1 in
Height:	8 ft 10 in
Max loaded weight:	12,790 lb

First of two prototypes flew on 29 December 1948. Subsequently first swept wing aircraft to land on a carrier. Not developed for service use.

De Havilland Sea Venom FAW.21
Two-seat all-weather jet fighter

Power:	One de Havilland Ghost 104 turbojet, 4,950 lb st
Armament:	Four 20 mm cannon
Max speed:	630 mph
Range:	1,000 miles
Initial rate of climb:	8,760 ft/min
Span:	42 ft 11 in
Length:	36 ft 7 in
Height:	6 ft 6 in
Empty weight:	11,000 lb
Loaded weight:	15,800 lb (max)

First flight on 19 April 1951. 256 Sea Venoms produced. In Royal Navy service 1954–60.

Supermarine Type 508
Single-seat twin-engined jet fighter

Power:	Two Rolls-Royce Avon R.A.3 axial flow turbojets, each 6,500 lb st
Armament:	Provision for four 30 mm cannon
Max speed:	590 mph at sea level
Range:	800 miles
Service ceiling:	45,000 ft
Span:	41 ft (20 ft folded)

Length:	46 ft 9 in
Height:	11 ft 7 in
Empty weight:	18,000 lb

First prototype flew on 31 August 1951. Second prototype completed as similar Type 529. Further development led to Type 525 and Scimitar (see below).

De Havilland DH.110 Sea Vixen FAW.1
Two-seat all-weather twin-engined jet fighter

Power:	Two Rolls-Royce Avon 208 axial flow turbojets, each 11,230 lb st
Armament:	Four Firestreak air-to-air guided missiles plus 28 2-in unguided rockets; two 1,000 lb bombs instead of missiles
Max speed:	645 mph at 10,000 ft
Initial rate of climb:	10,000 ft/min
Service ceiling:	48,000 ft
Span:	50 ft
Length:	55 ft 7 in
Height:	10 ft 9 in
Loaded weight:	36,000 lb

DH.110 prototype flew on 26 September 1951. First navalised prototype flew on 20 June 1955. In service 1958–72.

Supermarine Type 544 Scimitar F.1
Single-seat twin-engined jet fighter-bomber

Power:	Two Rolls-Royce Avon 202 axial flow turbojets, each 11,250 lb st
Armament:	Four 30 mm cannon; provision for up to 4,000 of bombs, or four Bullpup air-to-surface missiles or four Sidewinder air-to-air missiles; also capable of carrying tactical nuclear bombs
Max speed:	710 mph at 10,000 ft
Range:	1,422 miles
Initial rate of climb:	12,000 ft/min
Service ceiling:	46,000 ft

Span:	37 ft 2 in (20 ft 6 in folded)
Length:	55 ft 4 in
Height:	15 ft 3 in
Empty weight:	23,960 lb
Loaded weight:	34,200 lb (max)

Prototype Type 525 flew on 27 April 1954. First Type 544 Scimitar flew on 20 January 1956. 76 production examples. In service 1957–66.

UNITED STATES NAVY JET AIRCRAFT

Fighters/Fighter-bombers

Ryan FR-1 Fireball
Single-seat composite-powered fighter

Power:	One Wright R-1820-72W radial piston engine, 1,350 hp, and one General Electric J31 turbojet, 1,600 lb thrust
Armament:	Four 0.5-in machine-guns
Max speed:	404 mph (both engines) at 17,800 ft
Service ceiling:	43,100 ft
Range:	1,620 st. miles
Span:	40 ft 0 in
Length:	32 ft 4 in
Height:	13 ft 11 in
Empty weight:	7,689 lb
Loaded weight:	11,652 lb

Ordered in February 1943. First flight on 25 June 1944. Three prototypes and 66 production examples built. In service 1945–7.

McDonnell FH-1 Phantom
Single-seat twin-jet fighter

Power:	Two Westinghouse J30-WE-20 axial flow turbojets, each 1,600 lb st
Armament:	Four nose-mounted Browning 0.5-in machine-guns
Max speed:	479 mph at sea level
Range:	695 miles at 250 mph

Initial rate of climb: 4,230 ft/min
Service ceiling: 41,000 ft
Span: 40 ft 9 in
Length: 38 ft 9 in
Height: 14 ft 2 in
Empty weight: 6,683 lb
Loaded weight: 12,035 lb (max)

Prototype XFD-1 first flew on 26 January 1945. 2 prototypes and 60 production FH-1s. In service 1947–53.

Curtiss XF15C-1
Single-seat composite-powered fighter

Power: One Pratt & Whitney R-2800-34W radial piston engine, 2,100 hp, and one Allis-Chalmers/de Havilland H1-B Goblin turbojet, 2,700 lb thrust
Armament: Four 20 mm cannon
Max speed: 469 mph (both engines) at 25,300 ft
Initial climb rate: 5,020 ft/min (both engines)
Service ceiling: 41,800 ft
Range: 1,385 miles
Span: 48 ft 0 in
Length: 44 ft 0 in
Height: 15 ft 3 in
Empty weight: 12,648 lb
Loaded weight: 16,630 lb

Ordered in April 1944. First flight on 27 February 1945. Three prototypes built, one of which crashed in May 1945.

Chance Vought F6U-1 Pirate
Single-seat jet fighter

Power: One Westinghouse J34-WE-30 turbojet with afterburning. 3,150 lb dry thrust
Armament: Four 20 mm cannon
Max speed: 596 mph at sea level
Range (max with tip tanks):
                          1,170 miles

Initial rate of climb: 8,000 ft/min
Service ceiling: 46,000 ft
Span: 32 ft 10 in
Length: 37 ft 7 in
Height: 12 ft 11 in
Empty weight: 7,320 lb
Loaded weight (max): 12,900 lb

Prototype XF6U-1 flew on 2 October 1946. 30 production examples used for development purposes. No operational service.

North American FJ-1 Fury
Single-seat jet fighter

Power: One Allison J35-A-2 turbojet,
                              4,000 lb st
Armament: Six 0.5-in machine-guns
Max speed: 547 mph at 9,000 ft
Range (ferry):   1,500 miles
Initial rate of climb: 3,300 ft/min
Service ceiling: 32,000 ft
Span: 38 ft 2 in
Length: 34 ft 5 in
Height: 14 ft 10 in
Empty weight: 8,843 lb
Loaded weight (max): 15,600 lb

Prototype XFJ-1 flew on 27 November 1946. 3 prototypes and 30 production aircraft. Equipped world's first fully operational carrier-based squadron. In frontline service 1948–9.

McDonnell F2H-2 Banshee
Single-seat twin jet fighter

Power: Two Westinghouse J34-WE-34 axial flow
                              turbojets, each 3,250 lb st
Armament: Four 20 mm cannon
Max speed: 532 mph at 10,000 ft
Range: 1,475 miles
Initial rate of climb: 3,910 ft/min
Service ceiling: 44,800 ft

Span:	44 ft 10 in
Length:	40 ft 2 in
Height:	14 ft 6 in
Empty weight:	11,145 lb
Loaded weight:	22,312 lb

Prototype first flew on 11 January 1947 as XF2D-1. 882 Banshees delivered. In US Navy service 1948–62.

Grumman F9F-2 Panther
Single-seat jet fighter

Power:	One Pratt & Whitney J42-P-6 turbojet, 5,000 lb st
Armament:	Four 20 mm cannon; up to 3,000 lb of ordnance stores
Max speed:	594 mph at sea level
Range:	1,100 miles
Initial rate of climb:	7,700 ft/min
Service ceiling:	43,000 ft
Span:	35 ft 3 in (23 ft 5 in folded)
Length:	37 ft 8 in
Height:	11 ft 3 in
Empty weight:	7,107 lb
Loaded weight (max):	12,442 lb

Prototype XF9F-2 flew on 24 November 1947. 1,387 Panthers produced including prototypes. In US Navy/USMC service 1949–58.

Douglas F3D-1 Skyknight
Two-seat twin jet night fighter

Power:	Two Westinghouse J34-WE-36 turbojets, each 3,400 lb st
Armament:	Four 20 mm cannon
Max speed:	565 mph at 20,000 ft
Range (max):	1,540 miles
Initial rate of climb:	4,500 ft/min
Service ceiling:	38,200 ft
Span:	50 ft

Length:	45 ft 5 in
Height:	16 ft 1 in
Empty weight:	18,160 lb
Loaded (max):	27,680 lb

Prototype XF3D-1 flew on 23 March 1948. 266 produced including three prototypes and 28 F3D-1s. In service 1951–70. Mostly used by shore-based USMC units.

Chance Vought F7U-3 Cutlass
Single-seat twin engine tailless jet fighter

Power:	Two Westinghouse J46-WE-8A turbojets with afterburning, 4,600 lb dry thrust
Armament:	Four 20 mm cannon; later provision for four Sparrow air-to-air missiles
Max speed:	680 mph at 10,000 ft
Range:	660 miles
Initial rate of climb:	13,000 ft/min
Service ceiling:	40,000 ft
Span:	38 ft 8 in
Length:	44 ft 3 in
Height:	14 ft 7 in
Empty weight:	18,210 lb
Loaded weight (max):	31,642 lb

Prototype XF7U-1 flew on 1 March 1950. 290 of all variants produced. In US Navy service 1952–9.

Douglas F4D-1 Skyray
Single-seat jet interceptor

Power:	One Pratt & Whitney J57-P-2 turbojet, 9,700 lb st (14,500 lb st with afterburning)
Armament:	Four 20 mm cannon; up to 4,000 lb of ordnance stores including Sidewinder air-to-air missiles
Max speed:	695 mph at 36,000 ft
Range (max):	1,200 miles
Initial rate of climb:	18,000 ft/min
Service ceiling:	55,000 ft

Span:	33 ft 6 in
Length:	45 ft 8 in
Height:	13 ft
Empty weight:	16,025 lb
Loaded (max):	25,000 lb

Prototype XF4D-1 flew on 23 January 1951. 420 delivered. In service 1956–64. US Navy's first fighter capable of exceeding Mach 1 in level flight.

McDonnell F3H-2 Demon
Single-seat jet fighter

Power:	One Allison J71-A-2E turbojet, 9,700 lb st
Armament:	Four 20 mm cannon; later versions had provision to carry four Sidewinder air-to-air missiles or up to 6,000 lb of ordnance stores
Max speed:	647 mph at 30,000 ft
Range:	1,370 miles
Initial rate of climb:	12,795 ft/min
Service ceiling:	42,650 ft
Span:	35 ft 4 in
Length:	58 ft 11 in
Height:	14 ft 7 in
Empty weight:	22,133 lb
Loaded weight (max):	33,900 lb

Prototype flew on 7 August 1951. 515 delivered. In service 1956–64.

Grumman F9F-6 Cougar
Single-seat jet fighter

Power:	One Pratt & Whitney J48-P-8 turbojet, 7,250 lb st
Armament:	Four 20 mm cannon; two 1,000 lb bombs on underwing racks
Max speed:	654 mph at sea level
Range:	932 miles
Initial rate of climb:	6,750 ft/min

Service ceiling: 44,600 ft
Span: 36 ft 5 in (14 ft 2 in folded)
Length: 41 ft 7 in
Height: 15 ft
Empty weight: 11,255 lb
Loaded weight (max): 21,000 lb

Prototype XF9F-6 flew on 20 September 1951. 1,988 Cougars produced including prototypes. In frontline US Navy service 1952–60.

North American FJ-2 Fury
Single-seat swept-wing jet fighter

Power: One General Electric J47-GE-2 turbojet, 6,000 lb st
Armament: Four 20 mm cannon
Max speed: 677 mph at sea level
Range: 990 miles
Initial rate of climb: 7,230 ft/min
Service ceiling: 41,700 ft
Span: 37 ft 1 in
Length: 37 ft 7 in
Height: 13 ft 7 in
Empty weight: 11,800 lb
Loaded weight (max): 18,790 lb

Naval version of F-86E Sabre. Prototype XFJ-2B flew on 27 December 1951. 1,115 FJ-2/3/4s produced. In service 1953–63.

Grumman XF10F-1 Jaguar
Single-seat variable geometry jet fighter

Power: One Westinghouse XJ40-WE-6, 6,800 lb dry thrust
Armament: Provision for four 20 mm cannon; up to 2,000 lb of bombs or rockets on underwing racks
Projected performance:
Max speed: 710 mph at sea level
Range: 1,670 miles

Initial rate of climb: 13,350 ft/min (with afterburning)
Service ceiling: 45,800 ft
Span: 50 ft 7 in (minimum sweepback),
                          24 ft 9 in folded
Length: 54 ft 5 in
Height: 16 ft 3 in
Empty weight: 20,426 lb
Loaded weight (max): 35,450 lb

Prototype flew on 19 May 1952. Second prototype completed but not flown. Project cancelled in April 1953.

Convair XF2Y-1 Sea Dart
Single-seat twin jet seaplane fighter

Power: Two Westinghouse J34-WE-32 turbojets,
                          each 3,400 lb st
Armament: None fitted
Max speed: 690 mph at sea level
Span: 30 ft 6 in
Length: 41 ft 2 in
Height: 21 ft 1 in (with hydro-ski extended)
Loaded weight: 22,000 lb

Prototype flew on 9 April 1953. Two XF2Y-1s and two YF2F-1s produced. Programme cancelled in 1956.

Grumman F11F-1 Tiger
Single-seat jet fighter

Power: One Wright J65-W-18 turbojet,
                          7,450 lb st
Armament: Four 20 mm cannon; up to four
                          Sidewinder air-to-air missiles
Max speed: 750 mph at sea level
Range: 1,270 miles
Initial rate of climb: 5,130 ft/min
Service ceiling: 41,900 ft
Span: 31 ft 7 in (27 ft 4 in folded)
Length: 46 ft 11 in
Height: 13 ft 3 in

Empty weight: 13,307 lb
Loaded weight (max): 23,459 lb

Prototype XF11F-1 flew on 30 July 1954. 201 Tigers produced including prototypes. In frontline US Navy service 1957–61.

Chance Vought F8U-1 Crusader
Single-seat supersonic jet fighter

Power:	One Pratt & Whitney J57-P-11 turbojet with afterburning, 18,900 lb dry thrust
Armament:	Four 20 mm cannon; 32 air-to-air unguided rockets; provision for two Sidewinder AAMs
Max speed:	1,100 mph at 40,000 ft
Range:	1,200 miles
Service ceiling:	60,000 ft
Span:	35 ft 8 in
Length:	54 ft 3 in
Height:	15 ft 9in
Loaded weight (max):	27,000 lb

Prototype XF8U-1 flew on 25 March 1955. 1,291 Crusaders of all variants produced (including 42 for French *Aeronavale*). In US Navy service 1957–73.

Attack and Strike Aircraft

Douglas A3D-1 Skywarrior
Three-seat twin jet attack bomber

Power:	Two Pratt & Whitney J57-P-10 turbojets, each 12,400 lb st
Armament:	Two 20 mm cannon in remote-controlled rear barbette; up to 12,000 lb of bombs in internal bomb bay
Max speed:	610 mph at 10,000 ft
Range (max):	2,100 miles
Service ceiling:	41,000 ft
Span:	72 ft 6 in
Length:	76 ft 4 in

Height: 22 ft 9 in
Empty weight: 39,400 lb
Loaded weight (max): 82,000 lb

Prototype flew on 28 October 1952. 282 of all variants delivered. In US Navy service 1956–88.

Douglas A4D-1 Skyhawk
Single-seat jet attack bomber

Power: One Wright J65-2-4 turbojet, 7,700 lb st
Armament: Two 20 mm cannon; up to 5,000 lb of ordnance stores
Max speed: 664 mph at sea level
Range (max):   1,150 miles
Span: 27 ft 6 in
Length: 39 ft 5 in
Height: 15 ft
Empty weight: 8,400 lb
Loaded weight (max): 20,000 lb

Prototype XA4D-1 flew on 22 June 1954. Subsequently 2,960 Skyhawks of all variants delivered. Entered US Navy service in 1956.

Appendix IV

BRITISH AND AMERICAN AIRCRAFT CARRIERS – 1945 TO 1955

This appendix gives brief details of the main aircraft carrier classes that were in service between 1945 and 1955. Included are the projected British Malta class and the American USS *United States*. Dates in parenthesis indicate the year the ship was completed. Dimensions are to the nearest foot (ft). The following abbreviations are used: oa – overall; wl – waterline; pp – between perpendiculars.

ROYAL NAVY CARRIERS

Illustrious class fleet carriers (4 ships)

Data:	HMS *Illustrious* 1945
Displacement:	23,200 tons standard; 31,630 tons full load
Dimensions:	Length: 748 ft oa; 673 ft pp. Beam: 96 ft at waterline. Draught: 29 ft full load
Flight deck:	Length 740 ft; width 95 ft
Flight deck protection:	3-in armour over hangars, 1.5-in elsewhere
Armament:	16 4.5-in DP guns in eight twin turrets; 40 2-pdr guns in five octuple mountings, three single 40 mm, and 52 20 mm light AA guns
Aircraft:	54 maximum; one BH3 hydraulic catapult

Machinery:	Six Admiralty three-drum boilers, Parsons SR turbines, three shafts; 111,000 shp; 30 kt
Oil fuel:	4,850 tons; endurance 6,300 nm at 25 kt
Complement:	1,274; wartime 1,997
Notes:	*Illustrious* (1940), *Victorious* (1940), *Formidable* (1941) all built to the same design with a single hangar deck. *Indomitable* (1941) completed to a modified design with an extra lower half length hangar deck incorporated.

Implacable class fleet carriers (2 ships)

Data:	HMS *Implacable* 1945
Displacement:	27,000 tons standard; 32,110 tons full load
Dimensions:	Length: 766 ft oa. Beam: 96 ft at waterline. Draught: 29 ft full load
Flight deck:	Length 761 ft; width 101 ft
Flight deck protection:	3-in armour over hangars, 1.5-in elsewhere
Armament:	16 4.5-in DP guns in eight twin turrets; 48 2-pdr guns in six octuple mountings, 37 20 mm light AA guns
Aircraft:	81 maximum; one BH3 hydraulic catapult
Machinery:	Eight Admiralty three-drum boilers, Parsons SR turbines, four shafts; 148,000 shp; 30.5 kt at full load
Oil fuel:	4,850 tons
Complement:	1,492; wartime 1,574
Notes:	*Implacable* (1944), *Indefatigable* (1944). Development of Illustrious class with extended lower hangar and four-shaft machinery.

Illustrious class fleet carrier modernisation (1 ship)

| Data: | HMS *Victorious* 1957 as rebuilt |
| Displacement: | 30,530 tons standard; 35,500 tons full load |

Dimensions:	Length: 781 ft oa; 740 ft wl.
	Beam: 103 ft wl. Draught: 31 ft full load
Flight deck:	Length 775 ft; width 147 ft
Flight deck protection:	3-in armour over hangars, 1.5-in elsewhere
Armament:	12 3-in/50 cal in twin mountings. One Mk.6 sextuple 40 mm AA mounting
Aircraft:	28 fixed wing, 8 ASW helicopters
Machinery:	Six Foster Wheeler boilers, Parsons geared turbines, three shafts; 111,000 shp; 31 kt
Oil fuel:	4,850 tons
Complement:	2,400
Notes:	*Victorious* was the only Illustrious class ship to be rebuilt. Modernisation refit took seven years from 1950 to the end of 1957. An 8.75-degree angled deck was fitted together with two steam catapults and two mirror deck landing sights.

Audacious class fleet carriers (2 ships completed)

Data:	HMS *Eagle* (as completed 1951)
Displacement:	36,800 tons standard; 45,720 tons full load
Dimensions:	Length: 804 ft oa; 750 ft wl. Beam: 113 ft at waterline. Draught: 36 ft full load
Flight deck:	Length 800 ft; width 112 ft
Flight deck protection:	4-in armour over hangars, 1.5-in elsewhere
Armament:	16 4.5-in DP guns in eight twin turrets; 61 40 mm guns in eight Mk.6 sextuple, two Mk.5 twin and nine single mountings
Aircraft:	80 maximum; two BH5 hydraulic catapults
Machinery:	Eight Admiralty three-drum boilers, Parsons SR turbines, four shafts; 152,000 shp; 30.5 kt at full load

Oil fuel: 5,500 tons; endurance 5,000 nm at 24 kt
Complement: 2,250; wartime 2,750
Notes: Two ships, *Eagle* and *Ark Royal*. Latter
                        completed to revised design in 1955
                        with angled deck, deck edge lift,
                        steam catapults and 40 mm battery
                        reduced to 40 guns. *Eagle* fully
                        modernised in a refit bewteen 1959
                        and 1964.

Malta class fleet carriers (3 ships projected)

Data: HMS *Malta* as projected 1945
Displacement: 46,900 tons standard; 56,800 tons full
                        load
Dimensions: Length: 916 ft oa; 820 ft pp. Beam:
                        116 ft at waterline. Draught: 34 ft
                        full load
Flight deck: Length 909 ft; width 136 ft
Flight deck protection: 1-in armour plating
Armament: 16 4.5-in DP guns in eight twin turrets;
                        55 40 mm AA guns
Aircraft: 80 in hangar, over 100 including
                        deck park; two BH5 hydraulic
                        catapults
Machinery: Eight Admiralty three-drum boilers,
                        Parsons SR geared turbines, four shafts;
                        200,000 shp; 32 kt at full load
Oil fuel: 6,000 tons
Complement: 3,520
Notes: Three ships projected (*Malta*, *New
                        Zealand*, *Gibraltar*) but only *Malta* laid
                        down before whole programme was
                        cancelled in early 1946.

Colossus class light fleet carriers (10 ships)

Data: HMS *Vengeance* (as completed 1945)
Displacement: 13,190 tons standard; 18,040 tons full
                        load

Dimensions: Length: 694 ft oa. Beam: 80 ft at
                         waterline. Draught: 23 ft full load
Flight deck: Length 690 ft; width 80 ft
Flight deck protection: Unarmoured steel deck
Armament: 24 2-pdr guns in six quadruple
                         mountings; 32 20 mm AA guns in 11
                         twin and ten single mountings
Aircraft: 42 maximum; two BH3 hydraulic
                         catapults
Machinery: Four Admiralty three-drum boilers,
                         Parsons SR geared turbines, two shafts;
                         36,000 shp; 25 kt
Oil fuel: 3,196 tons; endurance 5,900 nm at
                         25 kt
Complement: Wartime 1,300
Notes: *Colossus* (1944), *Ocean* (1945), *Vengeance*
                         (1945), *Glory* (1945), *Venerable* (1945),
                         Warrior (1946), *Theseus* (1946), *Triumph*
                         (1946). *Perseus* and *Pioneer* both
                         completed in 1945 as Aircraft
                         Maintenance Ships.

Majestic class light fleet carriers (6 ships)
Data: Majestic class as designed
Displacement: 15,700 tons standard; 19,500 tons full
                         load
Dimensions: Length: 695 ft oa; 630 ft pp. Beam: 80 ft
                         at waterline. Draught: 26 ft full load
Flight deck: Length 690 ft; width 106 ft
Flight deck protection: Unarmoured steel deck
Armament: 32 40 mm in six twin and 20 single
                         mountings
Aircraft: 37 maximum; two BH3 hydraulic
                         catapults
Machinery: Four Admiralty three-drum boilers,
                         Parsons SR geared turbines, two shafts;
                         36,000 shp; 25 kt
Oil fuel: 3,196 tons; endurance 6,200 nm at 23 kt
Complement: 1,100; wartime 1,343

Notes: None served with Royal Navy. *Terrible*,
                        completed in 1949 as HMAS *Sydney*,
                        and *Magnificant* (1948), loaned to the
                        RCN, were both built to the original
                        configuration. *Powerful* (1957) completed
                        as HMCS *Bonaventure*, *Majestic* (1955)
                        completed as HMAS *Melbourne* and
                        Hercules (1961) completed as Indian
                        Navy Ship *Vikrant*, all incorporating
                        angled decks, steam catapults and
                        mirror landing sights. *Leviathan* was
                        laid up after being launched in 1945
                        and was never completed.

Centaur class light fleet carriers (4 ships)

Data: HMS *Centaur* as completed in 1953
Displacement: 20,260 tons standard; 27,800 tons full
                             load
Dimensions: Length: 737 ft oa; 686 ft wl. Beam: 90 ft
                             at waterline. Draught: 28 ft full load
Flight deck: Length 733 ft; width 103 ft
Flight deck protection: Unarmoured steel deck
Armament: 32 40 mm in two six-barrelled Mk.6,
                             eight Mk.5 twin and four Mk.7 single
                             mountings
Aircraft: 42 maximum; two BH5 hydraulic
                             catapults
Machinery: Four Admiralty three-drum boilers,
                             Parsons SR geared turbines, two shafts;
                             76,000shp; 27.25 kt
Oil fuel: 3,500 tons; endurance 5,650 nm at 20 kt
Complement: 1,400
Notes: *Centaur* completed as designed in 1953.
                             Albion and *Bulwark* completed in 1954
                             with interim 5.75-degree angled deck.
                             Fourth ship, HMS *Hermes*, completed
                             to a revised design in 1959 and was
                             equipped with fully angled deck,
                             steam catapults and deck edge lift.

US NAVY CARRIERS

Essex class fleet carriers (23 ships)

Data:	USS *Leyte* as completed in 1946
Displacement:	27,100 tons standard; 33,000 tons full load
Dimensions:	Length: 888 ft oa. Beam: 93 ft wl. Draught: 28 ft full load
Flight deck:	Length 860 ft; width 96 ft
Flight deck protection:	Unarmoured steel deck overlaid with wood decking
Armament:	Twelve 5-in/38 DP guns in four twin and four single mountings; 44 40 mm and 19 20 mm AA guns
Aircraft:	100 maximum; two H-4 hydraulic catapults
Machinery:	Eight Babcock & Wilcox boilers, Westinghouse geared turbines, four shafts; 150,000 shp; 33 kt
Oil fuel:	6,161 tons; endurance 14,100 nm at 20 kt
Complement:	3,448
Notes:	CV-9 *Essex* (1942). CV-10 *Yorktown*, CV-11 *Intrepid*, CV-12 *Hornet*, CV-16 *Lexington*, CV-17 *Bunker Hill*, CV-18 *Wasp* (1943). CV-13 *Franklin*, CV-14 *Ticonderoga*, CV-15 *Randolph*, CCV-18 *Hancock*, CV-20 *Bennington*, CV-31 *Bon Homme Richard*, CV-38 *Shangri La* (1944). CV-21 *Boxer*, CV-36 *Antietam*, CV-37 *Princeton*, CV-39 *Lake Champlain*, CV-40 *Tarawa* (1945). CV-32 *Leyte*, CV-33 *Kearsarge*, CV-45 *Valley Forge*, CV-47 *Philippine Sea* (1946). Most of these ships were substantially modified under the SCB-27A and -27C programmes (see Chapters 4 and 6).

Modified Essex class fleet carrier (1 ship)

Data:	USS *Oriskany* (CV-34) as completed 1950
Displacement:	30,800 tons standard; 39,800 tons full load
Dimensions:	Length: 888 ft oa. Beam: 129 ft over bulges. Draught: 28 ft full load
Flight deck:	Length 860 ft; width 147 ft
Flight deck protection:	Unarmoured steel deck
Armament:	Eight 5-in DP guns in single mountings; 28 3-in/50 cal AA guns in twin mountings
Aircraft:	80 maximum; two H-8 hydraulic catapults
Machinery:	Eight Babcock & Wilcox boilers, Westinghouse geared turbines, four shafts; 150,000 shp; 33 kt
Oil fuel:	6,161 tons; endurance 14,100 nm at 20 kt
Complement:	3,460
Notes:	Last Essex class carrier. Completed in 1950 to SCB-27A standard.

Midway class fleet carriers (3 ships)

Data:	USS *Midway* as completed in 1945
Displacement:	45,000 tons standard; 60,000 tons full load
Dimensions:	Length: 968 ft oa, 900 ft wl. Beam: 113 ft wl. Draught: 35 ft full load
Flight deck:	Length 932 ft; width 136 ft
Flight deck protection:	Up to 3.5-in armour plating
Armament:	Eighteen 5-in/54 cal DP guns in single mountings; 84 40 mm and 38 20 mm AA guns
Aircraft:	137 maximum; two H-4 hydraulic catapults
Machinery:	Twelve Babcock & Wilcox boilers, Westinghouse geared turbines, four shafts; 212,000shp; 33 kt

Oil fuel: 9,276 tons; endurance 9,600 nm at 20 kt
Complement: 4,104 (war)
Notes: *Midway* (CVB-41) and *Franklin*
                           D Roosevelt (CVB-42) completed in
                           1945. *Coral Sea* (CVB-43) completed in
                           1947. All three modified between 1954
                           and 1960 to incorporate fully angled
                           deck, C-11 steam catapults and revised
                           deck lift layout.

United States class 'super carrier' (1 ship projected)

Data: USS *United States* as projected in 1948
Displacement: 66,850 tons standard; 78,500 tons full
                           load
Dimensions: Length: 1,090 ft oa, 1,030 ft wl. Beam:
                           130 ft wl. Draught: 35 ft full load
Flight deck: Length 1,025 ft; width 190 ft
Flight deck protection: 3-in armour plating
Armament: Eight 5-in/54 cal DP guns in single
                           mountings; 12 3-in/70 cal AA guns in
                           twin mounts; 20 20 mm AA guns
Aircraft: 75 jet fighters and 18 nuclear bombers;
                           four C-10 explosive-charge catapults
Machinery: Eight boilers, geared turbines, four
                           shafts; 280,000 shp; 33 kts
Oil fuel: 11,505 tons; endurance 12,000 nm at 10 kt
Complement: 4,127
Notes: USS *United States* (CVB-58) was laid
                           down on 14 April 1949 but the project
                           was cancelled only nine days later!
                           However, much of the design work was
                           carried forward into the succeeding
                           Forrestal class (see below).

Forrestal class super carriers (4 ships)

Data: USS *Forrestal* (CVA-59) as completed in
                           1955
Displacement: 60,000 tons standard; 78,000 tons full
                           load

Dimensions: Length: 1,039 ft oa, 990 ft wl. Beam: 129 ft wl. Draught: 37 ft full load

Flight deck: Length 1,015 ft; width 240 ft

Flight deck protection: 3-in steel decking (only partly armoured)

Armament: Eight 5-in/54 cal DP guns in single mountings

Aircraft: 95 maximum; two C-11 (bow) and two C-7 (angled deck) steam catapults

Machinery: Eight Babcock & Wilcox boilers, geared turbines, four shafts; 260,000 shp; 33 kt

Oil fuel: 7,800 tons; endurance 12,000 nm at 20 kt

Complement: 4,142

Notes: CVA-59 *Forrestal* (1955), CVA-60 *Saratoga* (1956), CVA-61 *Ranger* (1957), CVA-61 *Independence* (1959). With a wide flight deck, angled deck, four steam catapults and four deck edge lifts, these ships set the standard for all subsequent US carriers, which retain the same basic layout and similar dimensions, although the current Nimitz class carriers are, of course, nuclear powered.

SELECTED BIBLIOGRAPHY

Aircraft Carriers of the US Navy, Stefan Terzibaschitsch, Conway Maritime Press (1980)

Aircraft Carriers of the World, 1914 to the Present, Roger Chesneau, Arms and Armour Press (1984)

British Naval Aircraft since 1912, Owen Thetford, Putnam (1978)

Carrier Air Power, Norman Friedman, Conway Maritime Press (1981)

Carrier Aviation. Air Power Directory, David Donald and Daniel J March, Airtime Publishing (2001)

Carrier Operations in World War II. Vol. 1, The Royal Navy, David Brown, Ian Allan Ltd (1974)

Combat Aircraft Prototypes since 1945, Robert Jackson, Airlife Publishing Ltd (1985)

Curtiss Aircraft 1907–1947, Peter M Bowers, Putnam (1979)

de Havilland Aircraft since 1909, AJ Jackson, Putnam, Revised Ed. (1987)

Fairey Aircraft since 1915, HA Taylor, Putnam (1988) (reprint)

Fly Navy. The View from the Cockpit 1945–2000, Charles Manning (ed), Leo Cooper (2000)

Grumman Aircraft since 1929, Rene J Francillon, Putnam (1989)

Hawker Aircraft since 1920, Francis K Mason, Putnam, Revised Ed. (1991)

Hawker Typhoon, Tempest and Sea Fury, Kev Darling, Crowood Press (2003)

Royal Navy Aircraft Carriers 1945–1990, Leo Marriot, Ian Allan Ltd (1985)

Spitfire. The True Story of a Famous Fighter, Bruce Robertson, Harleyford Publications (1960)

Supermarine Aircraft since 1914, CF Andrews and EB Morgan, Putnam (1987)

The Jet Aircraft of the World, William Green and Roy Cross, Macdonald (1956)

The Seafire. The Spitfire that went to Sea, David Brown, Greenhill Books (1989)

The Squadrons of the Fleet Air Arm, Ray Sturtivant, Air Britain (1984)

United States Naval Aircraft since 1911, Gordon Swanborough & Peter M Bowers, Putnam (1968)

Warplanes of the Second World War, Fighters Vols 2 and 4, William Green, MacDonald (1961)

INDEX